李家鼎 著

阿爺廚房

消失的味道

萬里機構

推薦序一

　　2016 年 12 月 12 日是《阿爺廚房》首播的日子，鼎爺（李家鼎）首次穿起廚師服和阿譚（譚玉瑛）一起主持烹飪節目。一夜間，鼎爺由一位演員，踏上了廚藝之路……

　　在我眼裏，這個不只是一個藝人成名的故事，而是一個香港人成功的傳奇故事。一個在電視圈默默耕耘四十多年的藝人，年屆七十，終於通過《阿爺廚房》這個節目，成功打開一片天！

　　《阿爺廚房》有幸得到觀眾的喜愛，台前幕後功不可沒。雖然台前只有鼎爺和阿譚二人，但幕後卻有二十人之多，我們上上下下都是最佳拍檔。當主持人負責用言語、表情和廚藝介紹每道菜式時，幕後工作人員就會用鏡頭、燈光、拍攝手法、配樂，還有佈景、道具、服裝等等去配合，透過細緻的表達方式，解構菜式的做法，讓觀眾去回味、去欣賞、去學習《阿爺廚房》為大家準備的傳統功夫菜！

　　這本《阿爺廚房 —— 消失的味道》紀錄了不少將近消失，甚至已消失的菜式，也有一些傳統菜的做法。套用鼎爺在節目中常用的一句話：「不用急，慢慢來！」任何功夫菜都不會再難倒大家！

張漢翹

《阿爺廚房》監製

推薦序二

　　家鼎哥在烹飪上的手藝，是自幼在父親嚴厲教導之下，一點一滴調教過來的傳統粵菜功夫，由家常小菜至宴客大菜，不少已經失傳，而且漸漸被人遺忘。

　　值得慶幸的是，多得家鼎哥，多得《阿爺廚房》，一些已經變得陌生的傳統菜式，再次在電視上出現。

　　家鼎哥在《阿爺廚房》節目入面，將烹調竅門和心得毫無保留和大家分享，欣賞他的廚藝也是賞心悅目，特別是他純熟的刀功和下廚的功架，以全他對食材的認識、烹飪技巧的見解，用心一看、用心一聽就知道一定下過不少苦功，就算演技多好的演員也不容易演到這種程度；所以實在令人佩服，亦好值得去學習。

　　《阿爺廚房——消失的味道》就是家鼎哥幾十年的經驗之談，非常值得參考！

薛家燕

推薦序三

　　鼎爺一向給人的印象是好「打」得，自從《阿爺廚房》播映，鼎爺有幾「煮」得，不用再多說了。

　　《阿爺廚房》的特色是懷舊和傳統，這亦是鼎爺的強項，若然有機會可以重現眼前，給年長一輩觀眾的是一種回味，給年輕一輩觀眾的是一種見識，在這個前設之下，《阿爺廚房》就在 2016 年的年底正式和大家見面。

　　許多已給人淡忘的功夫菜，再次出現在節目之中，牽起了不少話題，大家對功夫菜的興趣也回復過來。這次能夠透過節目的影響力，將傳統文化承傳下去，是《阿爺廚房》最大的收穫。

　　《阿爺廚房——消失的味道》一書精選了《阿爺廚房》第五輯的 42 道菜式，有些菜式拍攝時才第一次見識到，就如西瓜酸、鮑魚綠豆沙等等，原來存在已久，但漸漸為人所淡忘，這正是這本食譜難得之處，值得大家收藏。

林志榮

《阿爺廚房》監製

推 薦 序 四

嘩！開心嘅事當然要嗌三聲啦！好嘢、好嘢、好好嘢。

我所講嘅係《阿爺廚房》拍到第五輯所出版嘅呢本節目輯錄呀！

回想當初跟鼎爺見面只說 Hi、Bye，皆因從未合作過，直至現在拍攝中偶爾會搭膊頭，公餘還談天說地，由工作理念至生活態度，無所不談，這種熟稔令拍攝時更揮灑自如。

書中這種消失的味道，在這年頭更令人懷念，有歷練的前輩固然會回味，年輕的也可從中了解過往社會的變遷。

總感覺鼎爺能夠充分發展自己力量去生活，更顯意義！

譚玉瑛

推薦序五

兩個口水多過茶嘅人，當然一見如故。

就係因為我多嘢講，鼎爺又多嘢講，所以講下講下就變成無所不談。

喺鼎爺未拍飲食節目之前，已經親耳聽聞佢口述廚藝，由起初有少少懷疑，到愈聽愈對路，直到佢有一日話我知，佢會主持一個飲食節目，我梗係戥佢開心都嚟唔切啦！

鼎爺嘅武術、馬術頂呱呱好多人都知，但到七十歲竟然仲有廚藝呢瓣畀人發掘出嚟，《阿爺廚房》一拍就拍咗五輯，睇住佢愈煮愈有，愈煮愈旺，連私房菜都開埋，「廚藝可以改變命運」原來唔係冇可能，鼎爺做咗完美示範！

呢本《阿爺廚房 —— 消失的味道》咪就係秘笈囉！大家真係要跟鼎爺學返下功夫菜，下一個行運可能係你！

林盛斌（BOB）

推薦序六

好多人話，鼎爺行了尾運，突然七十歲紅起來！

是的，鼎爺真的有運氣，遇上《阿爺廚房》，人生的確不同了。

可是，如果只有運氣，沒有大家台前幕後的努力和實力，相信《阿爺廚房》只會是一個曇花一現的飲食節目，而不會是至今拍了超過一百集的《阿爺廚房》。

即使過往拍過多少飲食節目，拍《阿爺廚房》都要由零開始，因為節目要拍的是傳統功夫菜。菜式方面，既要用盡方法喚醒鼎爺昔日為父親下廚的記憶，也要翻查資料搜尋古老菜式及做法，了解菜式背後的故事。因此，我希望《阿爺廚房》不只是一個單純教做菜的烹飪節目，更期望它是一個可以帶出昔日社會風貌的美食紀錄。

多一點功夫，多一點執着，成就了《阿爺廚房 —— 消失的味道》！一本誠意十足的食譜，請大家好好細味！

馬嘉茵
《阿爺廚房》編審

序

當初拍攝《阿爺廚房》，只是一個 20 集的煮食節目，想不到一拍就是五輯，總共有 150 集了。

年幼時為嘴刁的父親下廚，得個「做」字，專心地做、用心地做，每次開飯總懷着戰戰兢兢的心情，從來不敢奢求父親讚好，但日子有功，竟然不知不覺累積了不少功夫；老來有機會在節目上再「做」一遍，尋回不少回憶，也得到大家欣賞，老懷安慰，相信在天上的父親大人，應該滿意！

以前的日子，食材款式不多，廚具種類有限，要滿足口腹之慾，唯有靠功夫手藝。今時今日，許多菜式失傳，死因是「功夫」沒了；因為功夫需要時間，時間就是金錢，人工也是金錢，要完全用以前那套做法，難度媲美拍功夫片。

《阿爺廚房》就不同了，菜式沒有成本計算，可以為所欲為，這次甚至將「消失的味道」重現，實在煮得非常過癮。

在此，感謝 TVB、感謝製作團隊、感謝我的拍檔譚玉瑛，感謝各位觀眾及讀者支持！希望大家喜歡《阿爺廚房 —— 消失的味道》。

作者簡介

李家鼎

李家鼎（鼎爺），香港著名演員、武術指導及馬術教練。

生於廣州富裕家庭，兒時住西關泮塘，父親李倬雲從事酒家、醬園、麵粉、絲綢生意，是正宗「西關大少」。

八歲的鼎爺，與家人逃難至香港，家道中落，但父親不改西關大少識飲識食的本色，鼎爺自少開始為父親下廚。

正所謂「食在廣州，廚出鳳城」，父親以前吃的主要是順德大廚的手藝，所以鼎爺的廚風深受順德菜的影響，透過嘴刁的父親口傳廚藝技巧，又自學鑽研刀工，希望菜式做得更精緻，更合父親的胃口。

2016 年，開始主持《阿爺廚房》，在節目中大展廚藝，精湛的廚藝、細膩的刀工，給大家認識一個全新的鼎爺，傳統功夫粵菜亦再次在香港牽起熱潮。

《阿爺廚房》拍攝花絮

　　在密鑼緊鼓的拍攝過程中，台前幕後各工作人員不辭勞苦、認真製作，希望帶來舊日懷舊菜式的點滴與情懷。

《阿爺廚房》台前幕後工作人員。

鼎爺與譚玉瑛第五度合作拍攝烹飪節目，大家默契十足，擦出火花。

惹味的懷舊菜式，佐以一口啤酒，爽呀！

每個環節製作認真，重現消失的味道。

CONTENTS

第一章
肉類

阿爺教你揀食材

　　要煮得一手好菜，怎少得挑選好食材呢？今次，就跟着我到街市、塔門走走，了解各樣材料的特色，再靈活運用及配搭於各款傳統菜裏。

滷水料

　　滷水材料視乎味道濃或淡而加入不同的香料。滷水料宜放在密封的容器內，擺放於陰涼處即可。

　　八角——要選大顆、厚身的，帶點啡紅色，但不要太深色。由於味道濃烈，配搭家禽及肉類最好。

　　草果——味道淡淡的，用來配魚、蝦、蟹等海鮮可去除腥味；如炮製羊腩煲時加一、兩粒草果進去，可去除羊肉的羶味，吃起來特別甘香嫩滑。

　　甘草——宜挑選外皮啡色的甘草。甘草片不能有洞，有洞代表被蟲蛀了。甘草味道甘甜、不太濃，配搭味道較濃的食材，可中和味道。

　　肉桂——味道較濃，適合炆、燉的菜式，是五香粉其中一款主要原料。

　　香葉——又稱為月桂葉，有一種芳香的味道，可以加進炆餸或煮湯內，但煮好後必須取出，而且不要放太多，會帶有苦味。

　　花椒——味道有點濃烈，吃起來舌頭有點麻，加上其他配料做菜很好吃；但不宜選太紅的，因為有可能是染色的花椒。

　　丁香——選大粒、乾身的，摸上手有點油潤的感覺。它的味道濃烈，煮好菜後要把它挑出來。

瀨尿蝦乾

　　瀨尿蝦乾呈棕色部分的是「春」，有膏的瀨尿蝦是雌性，吃起來較甘香。有些瀨尿蝦乾是生剝生曬，曬兩天就可以了。蒸熟的瀨尿蝦乾直接淋上生抽和熟油，鮮味十足，伴飯吃一流！

陳皮

　　我喜歡用陳皮入饌，例如牛肉餅、陳皮燉鴨、煮糖水等，選用曬製五年或以上的陳皮，有陳皮香及售價相宜。一般人會將陳皮的果瓤刮掉，其實內瓤有軟化頑痰的效用，所以不宜除去。

紮仔菜

　　與大頭菜同科，水上人煮餸通常會用上紮仔菜。醃好後，為了妥善保存，將大頭菜切好後再紮在一起，成為醃菜的紮仔菜。烹調時通常連葉一起使用，取其鹹香之味，汆水後下白鑊烘乾，可以減少鹹味及苦澀味。

中國乾葱與泰國乾葱

我煮菜時習慣使用大量乾葱，夠惹味！大家有留意到街市有不同種類的乾葱出售？有些是一大束紮起來，有些是裝在尼龍網袋裏。

原來紮在一起的乾葱多數來自中國，當地的農民將紅葱拿去曬，曬乾之後就束起來，掛晾着風乾，所以是一束束的。這些乾葱體型細小，但葱味較濃烈，味道很嗆，令菜式非常惹味。

裝在尼龍網袋的乾葱，大部分來自泰國，外形較中國乾葱大，氣味較淡，含水量較高。

蟬蛻

又名蟬衣，是蟬科昆蟲蚱蟬羽化後的蛻殼；每年初夏之時，在山東、河南、河北產量最多。蟬蛻含有豐富的蛋白質、氨基酸及甲殼素，對小孩子尤其有益，具祛風、止痛、止癢、清熱的作用。

將蟬蛻與牛腩同炆，可以令牛腩更滑、更軟、更香，同時不會破壞牛肉的味道。

豬網油

豬網油是豬胃部及橫膈膜之間的一層網狀脂肪，是以前傳統菜式使用的材料，潮州人會用來包鱔，可以去掉淡水魚的泥味，吃到魚的鮮味。

豬網油很難處理，要細心地慢慢原幅修剪出來，不要有太多破洞或缺口，一般可向相熟的肉檔預訂。將豬網油泡在加了玫瑰露酒的水裏，可去除異味。

排骨

排骨分為很多種類，有腓排、一字排、腩排等。

腓排是豬頸脊骨下近肩胛骨之排骨，每邊只有一塊，肉質偏瘦、柔軟嫩滑。

一字排是排骨中最美味的部分，只有三成是肥肉，無論蒸或其他烹調方法都很好吃。

腩排是近腹腔部分連骨帶肉的排骨，肉較厚，當中有白色的軟骨。

 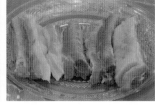 

腓排　　　　　　　　一字排　　　　　　　　腩排

海膽

香港出產的海膽於每年二月至六月當造，味道鮮甜肉厚。海膽和海刺很相似，一般人不容易分辨開來。其實細心地看，海膽的刺比較短，海刺的則較長，海刺味腥肉薄，味道不及海膽。

黃鱔

小時候，我為了吃黃鱔，曾在禾田裏的小水坑捉黃鱔，把手指放進鱔洞做餌，咬得傷痕纍纍。

對黃鱔的挑選及烹調，有以下少少心得：黃鱔要選粗大的，摸起來身上的肌肉結實，宰了後切片，炒起來鮮甜、有彈性、爽脆。

阿爺的入廚功夫

日常烹調做菜有特別的竅門，以下是我的入廚技巧，希望你們能夠從中領會，令廚藝精進。

醃味有先有後

我入廚的調味以簡單為主，但醃肉調味卻很有原則，先後有序。

我的基本調味方法是，先下糖以軟化食材；接着下油令質感更滑溜；加入生粉包裹食材，最後以生抽、鹽等滲入材料，但要緊記每下一種調味時要拌勻才下另一種；由淡味開始醃味，令食材得以好好地吸收每項調味之精華，達到增鮮之目的。

冰糖之妙用

　　冰糖除了用於調味之外，在炆煮肉類時，也會加入冰糖一起煮，能夠令肉質酥軟好吃。此外，今次於幾個菜式中，加入用冰糖煮成的焦糖溶液，令賣相呈現金黃色澤，更亮麗！

炸物時潠水

　　在炸豬油渣或魚乾時，於最後步驟階段潠入少許水，令炸物的口感更鬆脆；但切記進行時有一定危險程度，宜用鑊蓋輕蓋，以免熱油四濺燙傷，小心小心！

阿爺的自家製品

我對煮食有一份執着，對每樣配料都有要求，就算工夫較多的自家製配料，也樂意製作，為着的是對方吃進嘴裏滿足的一刻。

自家製沙茶醬

~Homemade Sa Cha sauce~

材料

蝦米	25 克
大地魚肉	25 克

辛香料

香葉	3 片
丁香	10 粒
五香粉	10 克
黃薑粉	10 克
辣椒粉	10 克
乾葱	190 克
葱	30 克
蒜頭	80 克

Ingredients

- 25 g dried shrimps
- 25 g dried plaice (de-boned, head, tail, fins removed)

Aromatics

- 3 bay leaves
- 10 cloves
- 10 g five-spice powder
- 10 g turmeric powder
- 10 g chilli powder
- 190 g shallots
- 30 g spring onion
- 80 g garlic

醬料

花生醬	4 湯匙
芝麻醬	2 湯匙
豆瓣醬	3 湯匙
蝦醬	1 湯匙
冰糖	100 克（舂碎）
蠔油	2 湯匙
油	300-400 毫升

Sauce

- 4 tbsp peanut butter
- 2 tbsp sesame paste
- 3 tbsp spicy soybean sauce
- 1 tbsp fermented shrimp paste
- 100 g rock sugar (crushed)
- 2 tbsp oyster sauce
- 300 to 400 ml oil

阿爺秘技

- 蝦米及大地魚肉攪打得愈幼細，沙茶醬會愈幼滑，味道互相融合。
- 炒沙茶醬必須不時注意火候，維持相同的溫度，並不斷地推勻，以免黏鍋炒焦。
- 製作沙茶醬需要不斷加入油分，炒料時若乾身可加入適量油。

Cooking tips

- The finer the dried shrimps and dried plaice are blended, the finer the Sa Cha sauce would be. The flavours also mingle and meld better that way.
- When you stir-fry the Sa Cha sauce, make sure you pay attention to the heat. Keep the wok at roughly the same temperature throughout and keep stirring all the time. Otherwise, the sauce may burn.
- Throughout the cooking process, you have to add oil from time to time whenever the mixture looks dry and starts to stick to the bottom of the wok.

做法

1..... 蝦米、大地魚肉放入白鑊烘香、脆口。
2..... 將蝦米、大地魚肉、香葉、丁香放入攪拌機打成粉狀，倒出。
3..... 將葱、乾葱、蒜頭放入攪拌機內打蓉。
4..... 起油鑊，倒入（3）炒香，加入（2）炒勻，推勻後調至中細火，加入少許油，依次加入五香粉、黃薑粉、辣椒粉推勻。
5..... 加入少許油，下花生醬、芝麻醬、豆瓣醬、蝦醬推勻，下冰糖、蠔油拌勻，熄火，不斷推勻，待涼，可放入已消毒之玻璃瓶儲存。

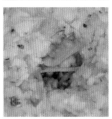

Method

1. Toast dried shrimps and dried plaice in a dry wok until crisp and lightly browned.
2. Put dried shrimps, dried plaice, bay leaves and cloves into a blender or food processor. Blend into a fine powder. Set aside.
3. Mince spring onion, shallots and garlic in a blender.
4. Heat wok and add oil. Stir-fry the garlic mixture from step 3 until fragrant. Add the dried shrimp powder from step 2. Stir to incorporate well and reduce to medium-low heat. Add a little of oil. Put in five-spice powder, turmeric powder and chilli powder in this particular order. Mix well.
5. Add a little of oil. Put in peanut butter, sesame paste, spicy soybean sauce and fermented shrimp paste. Stir to incorporate well. Add rock sugar and oyster sauce. Stir to mix well. Turn off the heat and keep stirring until the wok is cool. Transfer into a sterilized glass jar for storage.

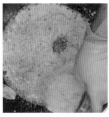

阿爺廚房——消失的味道

西 瓜 酸

~Watermelon rind pickles~

製 品 特 點

西瓜酸又叫西瓜仔，是海南一帶的特色食品，選取尚未成熟的小西瓜用粗鹽醃製。做好的西瓜酸甜中帶酸，開胃之餘還帶有西瓜的清香味。

【材料】

未成熟西瓜＿＿＿＿＿1 個
　（或大西瓜半個）
白醋＿＿＿＿＿＿＿1.5 公升
冰糖＿＿＿＿＿＿＿420 克
粗鹽＿＿＿＿＿＿＿適量
青檸＿＿＿＿＿＿＿1 個

Ingredients

- 1 mini unripe watermelon
  (or 1/2 of a large watermelon)
- 1.5 litres white vinegar
- 420 g rock sugar
- coarse salt
- 1 lime

【做法】

1..... 青檸切片，備用。

2..... 未成熟西瓜放入滾水，煮約 1 小時至皮軟身，取出，刺穿去掉
內部水分，加入粗鹽醃一日。

3..... 大西瓜切開，去掉全部瓜肉及綠色瓜皮，將剩下的西瓜青切成
粗塊，放入滾水汆燙半小時至軟，瀝乾水分，待涼，加粗鹽醃
製 1-2 日。

4..... 白醋加入冰糖煮至冰糖融化，放涼。

5..... 將迷你西瓜或大西瓜青的粗鹽抹走，放入已加熱消毒的玻璃瓶
內，加入糖醋水、青檸片及少許粗鹽，醃一星期便可食用。

阿 爺 秘 技 ——— Cooking tips

- 糖與醋的比例為 1 斤白醋、5 兩冰糖。
- 糖醋水煮沸後放涼才能使用，否則西瓜青被燙熟，難以
入味。
- 加入青檸片，令西瓜酸帶出獨特的香氣及酸味。
- 西瓜酸醃成後，汆水試味，覺得味道合適後可配其他材料
烹調。
- For the pickling brine, the ratio between white vinegar and
rock sugar is 600 ml to 188 g.
- The pickling brine should be left to cool completely
before poured over the watermelon rind. Otherwise, the
watermelon rind will be cooked by the brine and will not
pick up the flavours.
- The lime gives a characteristic citrus aroma and tartness to
the watermelon rind pickles.
- After the pickles are done, blanch them in boiling water and
taste them before serving. When the pickles are of the right
saltiness, you may use it to cook with other ingredients.

Method

1. Slice the lime.

2. If you use a mini unripe watermelon, cook it whole in boiling water for about 1 hour until the rind is softened. Drain. Pierce the rind to drain all water inside. Add coarse salt and rub evenly. Leave it for 1 day.

3. If you use half of a large watermelon, remove all red flesh and the dark green rind. Use only the white rind for this recipe. Cut the white rind into thick strips. Boil them in water for about 30 minutes until soft. Let cool. Add coarse salt and mix well. Leave them for 1 to 2 days.

4. To make the pickling brine, add rock sugar to white vinegar. Cook until sugar dissolves. Leave it to cool.

5. Pat off any salt on the watermelon rind. Put it into a sterilized glass jar. Pour in the pickling brine. Add sliced lime and a pinch of coarse salt. Leave it for 1 week. Serve.

▶ 示範短片

第一章

肉類

不論是排骨、牛腩，還是雞鴨等肉食，
在鼎爺的巧妙配搭之下，炮製出精彩絕
妙的特色懷舊菜，例如荷葉粉蒸肉、網
油腰肝拼臘腸卷、紙包雞、芋蓉鴨等，
還不快快拿起鑊鏟，依着步驟來做一頓
惹味肉食佳餚。

西瓜酸炒牛柳

~Stir-fried beef tenderloin with watermelon rind pickles~

菜式特點

用酸酸甜甜的西瓜酸來煮牛肉，不但沒蓋過牛肉的味道，而且連皮的西瓜酸咬起來還帶點嚼勁，兩者搭配別有一番風味！

【材料】

牛柳＿＿＿＿＿＿250 克
西瓜酸＿＿＿＿＿60 克
　（做法參考 p.23）
豆豉＿＿＿＿＿＿20 克
麻香 XXO 醬 ＿＿25 克
紅甜椒＿＿＿＿＿40 克
黃甜椒＿＿＿＿＿40 克
青甜椒＿＿＿＿＿40 克
葱段＿＿＿＿＿＿15 克
糖＿＿＿＿＿＿＿2 茶匙
生粉芡＿＿＿＿＿適量

Ingredients

- 250 g beef tenderloin
- 60 g watermelon rind pickles
  (see p.23 for recipe)
- 20 g fermented black beans
- 25 g XXO sauce with Sichuan pepper
- 40 g red bell pepper
- 40 g yellow bell pepper
- 40 g green bell pepper
- 15 g spring onion (cut into short lengths)
- 2 tsp sugar
- potato starch slurry

【醃料】

糖＿＿＿＿＿＿＿1 茶匙
生抽＿＿＿＿＿＿1 茶匙
油＿＿＿＿＿＿＿半茶匙
水＿＿＿＿＿＿＿半茶匙

Marinade

- 1 tsp sugar
- 1 tsp light soy sauce
- 1/2 tsp oil
- 1/2 tsp water

西瓜酸 Watermelon rind pickles

▶ 示範短片

【做法】

1..... 西瓜酸切條；豆豉切碎；牛柳切片；三色甜椒切條。
2..... 西瓜酸汆水，試味至味道合適後盛起備用。
3..... 牛柳下糖、生抽、油及水拌勻醃味。
4..... 起油鑊，下一半麻香XXO醬炒香，加入牛柳快炒至半熟，盛起備用。
5..... 起油鑊，加入西瓜酸、糖、豆豉、麻香XXO醬炒勻，加入三色甜椒翻炒，下牛柳炒勻，加入生粉芡，最後加入葱段炒勻即成。

Method

1. Cut watermelon rind pickles into strips. Finely chop the fermented black beans. Slice the beef. Cut all bell peppers into strips.
2. Blanch the watermelon rind pickles in boiling water. Taste it. Blanch further if it's still too salty. Drain and set aside.
3. Add sugar, light soy sauce, oil and water to the sliced beef. Mix well.
4. Heat wok and add oil. Add a half portion of XXO sauce with Sichuan pepper and stir-fry until fragrant. Add the sliced beef and stir-fry until half done. Set aside.
5. Heat wok and add oil. Stir-fry watermelon rind pickles, sugar, fermented black beans and XXO sauce with Sichuan pepper. Toss to mix well. Add bell peppers and toss again. Put in the beef and toss well. Pour in potato starch slurry. Add spring onion at last and toss. Serve.

阿爺秘技 ——————— Cooking tips

- 西瓜酸食用及烹調前必須汆水，以減少過鹹的味道。
- 烹調西瓜酸時，酌量加入糖以中和西瓜酸，更美味可口。
- 我喜歡用刀輕剁豆豉，以免豆豉舂得太爛，不香也沒咬口。

- Before serving watermelon rind pickles straight, or using them in cooking, make sure you blanch them in boiling water to remove the excessive saltiness.

- When you cook with watermelon rind pickles, season the dish with slightly more sugar to balance off the tartness and saltiness. The dish would taste better that way.

- I personally prefer to gently chop the fermented black beans with a knife, instead of pounding them with a knife handle. I find both their aroma and texture are lost when they are pounded too finely.

花膠雲吞

~Fish maw wontons~

菜式特點

花膠是用大魚的魚鰾曬乾製成的，基本可分為四大類：
廣肚、紮膠、鴨泡肚及鱔肚，這次選用較細小的鴨泡膠
代替雲吞皮製作「花膠雲吞」。

【雲吞材料】

110 頭鴨泡膠＿＿＿12 件
豬腩肉＿＿＿＿＿200 克
鮮蝦＿＿＿＿＿＿200 克
韭菜＿＿＿＿＿＿12 條
小棠菜＿＿＿＿＿4 棵（伴碟用）

【雲吞餡醃料】

生抽＿＿＿＿＿＿1 湯匙
大地魚粉＿＿＿＿3 湯匙
糖＿＿＿＿＿＿＿1 茶匙
鹽＿＿＿＿＿＿＿半茶匙
生粉＿＿＿＿＿＿1 茶匙

【上湯材料】

排骨＿＿＿＿＿＿600 克
雞＿＿＿＿＿＿＿1 隻
金華火腿＿＿＿＿50 克
雞腳＿＿＿＿＿＿10 隻
黃芽白＿＿＿＿＿300 克
水＿＿＿＿＿＿＿3 公升
蠔油＿＿＿＿＿＿2 湯匙（勾芡用）

Wontons

- 12 fish maws (110-head per 600 g)
- 200 g pork belly
- 200 g shrimps
- 12 sprigs Chinese chives
- 4 heads Shanghainese baby Bok Choy (as garnish)

Seasoning for filling

- 1 tbsp light soy sauce
- 3 tbsp ground plaice
- 1 tsp sugar
- 1/2 tsp salt
- 1 tsp potato starch

Stock

- 600 g pork ribs
- 1 chicken
- 50 g Jinhua ham
- 10 chicken feet
- 300 g Napa cabbage
- 3 litres water
- 2 tbsp oyster sauce (for thickening glaze)

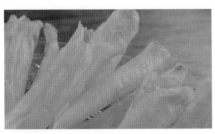

鴨泡膠 Fish maw

【做法】

1..... 鴨泡膠浸軟，洗淨後去掉雜質，備用。

2..... 鮮蝦去殼，洗淨、去腸，用毛巾索乾水分，用刀拍扁，再用刀背略剁成粒狀，製成有口感的蝦膠。

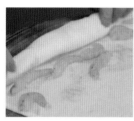

3..... 豬腩肉剁碎後，拌入蝦膠，加入醃料放雪櫃待 30 分鐘。

4..... 排骨、金華火腿、雞及雞腳全部汆水備用；黃芽白洗淨、切段。

5..... 湯鍋內注入 3 公升水，加入所有上湯材料，以大火煲滾後轉慢火煲 1-1.5 小時。

6..... 韭菜洗淨後，原條過熱水至軟身，瀝乾水分備用。

7..... 鍋內加水，水滾後將花膠置於蒸架上，大火蒸 5 分鐘至軟身，立即浸於室溫水降溫。

8..... 花膠抹乾，小心將醃好的雲吞餡釀進花膠內（釀進花膠中間位置），用韭菜把兩端紮起。

9..... 小棠菜洗淨，燙熟後過冷河，瀝乾水分，伴碟。

10... 將包好的花膠雲吞放碟上大火隔水蒸熟，淋上用上湯煮成的蠔油芡即可，或直接放進湯內進食。

Method

1. Soak fish maws in water till soft. Rinse and remove any dark bits.
2. Shell the shrimps. Rinse and devein. Wipe dry and crush with the flat side of a knife. Then coarsely chop them with the back of a knife. It's good to have some small chunks for extra mouthfeel.
3. Finely chop the pork belly. Stir in the chopped shrimp from step 2. Add seasoning and mix well. Leave the mixture in the fridge for 30 minutes.
4. To make the stock, blanch pork ribs, Jinhua ham, chicken and chicken feet together in boiling water. Drain and rinse well. Set aside. Rinse the Napa cabbage. Cut into short segments.
5. Pour 3 litres of water into a soup pot. Put in all stock ingredients. Bring to the boil over high heat. Turn to low heat and simmer for 1 to 1.5 hours.
6. Rinse the Chinese chives. Blanch them in boiling water until soft. Drain and set aside.
7. Put water into a wok or a steamer. Bring to the boil. Put in a steaming rack. Steam the fish maws over high heat for 5 minutes until soft. Soak in water at room temperature to cool them.
8. Drain and wipe dry fish maws. Carefully fill each fish maw with about 1 tsp of filling from step 3. Push the filling to the centre of the fish maw. Gather both ends of the fish maw and tie a sprig of Chinese chives to secure.
9. Rinse the Shanghainese baby Bok Choy. Blanch in boiling water and rinse in cold water. Drain and arrange along the rim of a serving plate.
10. Put the fish maw wontons on a steaming plate. Steam over high heat until cooked through. Serve in the stock directly from step 5. Or, heat a ladle of stock in a pan and stir in oyster sauce to make a glaze. Dribble the glaze over the steamed fish maw wontons and serve.

阿爺秘技 ——————— Cooking tips

- 經冷藏後的蝦肉，肉質爽口。
- 宜選用五花腩，肥瘦適中、油脂充足；而且花膠受油，腩肉的油分令花膠吸收油香，更香軟。
- 先用刀剟雞腳一下，出味之餘，也能充分釋出骨膠原。

- Shrimps turn springier in texture after refrigerated.
- For the filling, I use pork belly for the right proportion between fat and lean meat. As fish maws need to be cooked in fat to be tasty, the pork belly would complement them nicely with its fat to make them softer and more flavourful.
- I also cut the "palms" of chicken feet before making stock with them. That would let the flavours infuse in the stock and release the collagen therein.

沙茶醬牛柳炒芥蘭

~Stir-fried beef tenderloin with kale in Sa Cha sauce~

菜式特點

「芥蘭炒牛肉」是一道潮汕家常小菜，潮汕人會在這道菜加入沙茶醬，對於他們來說如沒有加入沙茶醬，這道炒牛肉就沒有靈魂了。

【材料】

牛柳	300 克
芥蘭	200 克
沙茶醬	70 克（做法參考 p.20）
紅尖椒	2 條（切圈）
糖	1 茶匙
蠔油	2 湯匙

【牛柳調味料】

蒜頭	2 粒（剁蓉）
乾葱	1 粒（剁蓉）
糖	1 茶匙
生抽	半茶匙
油	1 茶匙
水	2 湯匙

【炒芥蘭調味料】

糖	1 茶匙
紹興酒	1 茶匙

Ingredients

- 300 g beef tenderloin
- 200 g Chinese kale
- 70 g Sa Cha sauce (see p.20 for recipe)
- 2 red chillies (cut into rings)
- 1 tsp sugar
- 2 tbsp oyster sauce

Marinade for beef

- 2 cloves garlic (grated)
- 1 shallot (chopped)
- 1 tsp sugar
- 1/2 tsp light soy sauce
- 1 tsp oil
- 2 tbsp water

Seasoning for kale

- 1 tsp sugar
- 1 tsp Shaoxing wine

自家製沙茶醬
Homemade Sa Cha sauce

【做法】

1..... 牛柳切雙飛，下調味料拌勻。
2..... 芥蘭去掉菜葉，切莖部最嫩部分成蘭度。
3..... 起油鑊，下蘭度、糖、紹興酒快炒，盛起備用。
4..... 起油鑊，下部分沙茶醬及牛柳，炒至約五成熟，加
　　　入餘下沙茶醬及糖炒勻，下蘭度快炒，最後加入蠔
　　　油、紅椒圈炒勻即成。

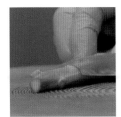

Method

1. Cut beef into butterflied slices. Add marinade and mix well.
2. Remove all leaves of the kale. Use only the most tender tips of each stem.
3. Heat wok and add oil. Put in the kale tips. Add sugar and Shaoxing wine. Toss quickly briefly. Set aside.
4. Heat the same wok again and add oil. Put in part of the Sa Cha sauce and sliced beef. Toss until beef is medium done. Put in the rest of the Sa Cha sauce and sugar. Toss to coat evenly. Add oyster sauce and red chillies at last. Toss and serve.

阿爺秘技 ———————————— Cooking tips

- 牛柳肉質嫩滑、無筋，適合配搭沙茶醬炒煮。
- 炒芥蘭時加入糖及紹興酒，可減少芥蘭的苦澀味。
- Beef tenderloin is the most tender cut without any sinews. It is the best choice for stir-frying in Sa Cha sauce.
- Adding sugar and Shaoxing wine to the kale when stir-frying helps remove its bitter taste.

話梅蒸豬脷

~Steamed pork tongue with liquorice plums~

菜式特點

這是一道農家菜，雖然豬脷不是很名貴的食材，不過意頭好，加上話梅鹹鹹酸酸的，是夏日的醒胃菜。

【材料】

甜話梅	6-8 粒
豬脷	400 克
紅椒絲	適量
芫荽	適量

Ingredients

- 6 to 8 dried liquorice plums
- 400 g pork tongue
- red chillies (finely shredded)
- coriander

【做法】

1..... 話梅用水浸軟，去核、切絲。
2..... 豬脷洗淨，氽水，切走表面的脷苔，斜切成片。
3..... 碟內放上豬脷片、話梅絲，蒸約 8 分鐘至熟透，
取出後放上紅椒絲、芫茜，淋上熟油即成。

Method

1. Soak dried plums in water till soft. De-seed and finely shred them.
2. Rinse pork tongue. Blanch in boiling water. Scrape off the coating on it. Slice diagonally.
3. Arrange pork tongue and shredded plums on a steaming plate. Steam for 8 minutes until cooked. Sprinkle with shredded red chillies and coriander on top. Drizzle with cooked oil. Serve.

阿爺秘技 ──────── Cooking tips

- 表面的脷苔必須切去，因它會有異味兼肉質粗糙。
- 挑選豬脷時，以舌底的肉呈粉紅色，舌面是白色，兩旁透出紅色的，摸起來軟軟的為佳。
- 豬脷不要切太薄，宜略有厚度，斜切成片除了快熟，吃起來也爽口。
- The coating on the pork tongue is quite gritty in texture and may carry an unpleasant smell. It must be thoroughly scraped off before used.
- When you shop for pork tongues, pick those with pink underside and white overside. The edges should look faintly red and it should feel soft when you press it gently.
- Do not slice the pork tongue too thinly. It gives a better mouthfeel when slightly on the thick side. Slicing it diagonally makes it cook faster, but also accentuates its springy texture.

荷葉粉蒸肉

~Steamed pork with ground spiced rice in lotus leaf~

菜式特點

「荷葉粉蒸肉」相傳與關羽的部將周倉有關。在一次出征途中，關羽讓周倉用荷葉把飯菜包起來，邊走邊吃，抄近路趕上隊伍。誰知，飯菜經荷葉一裹，散發出一股特有的芳香。後來經過一代代廚師不斷改進，演變成今天的菜式。

【材料】

乾荷葉＿＿＿＿＿2 塊
粘米＿＿＿＿＿150 克
糯米＿＿＿＿＿150 克
五花腩＿＿＿＿＿500 克
番薯＿＿＿＿＿200 克
八角＿＿＿＿＿2 粒
花椒＿＿＿＿＿1 湯匙
桂皮＿＿＿＿＿1 塊

【醃料】

五香粉＿＿＿＿＿1 湯匙
豆瓣醬＿＿＿＿＿3 湯匙
紹興酒＿＿＿＿＿2 湯匙
蒜蓉＿＿＿＿＿1 湯匙
乾葱蓉＿＿＿＿＿1 湯匙
糖＿＿＿＿＿2 茶匙
鹽＿＿＿＿＿1 茶匙
老抽＿＿＿＿＿1-2 茶匙

Ingredients

- 2 dried lotus leaves
- 150 g long-grain rice
- 150 g glutinous rice
- 500 g pork belly
- 200 g sweet potato
- 2 pods star-anise
- 1 tbsp Sichuan peppercorns
- 1 piece cassia bark

Marinade

- 1 tbsp five-spice powder
- 3 tbsp chilli bean sauce
- 2 tbsp Shaoxing wine
- 1 tbsp grated garlic
- 1 tbsp chopped shallot
- 2 tsp sugar
- 1 tsp salt
- 1 to 2 tsp dark soy sauce

【做法】

1..... 糯米用約 2 倍的水預先浸泡；乾荷葉用水浸軟。

2..... 五花腩切厚片；番薯切片。

3..... 五花腩依次加入醃料拌勻醃味。

4..... 白鑊放入糯米及粘米，加入八角、桂皮、花椒炒至乾身金黃，待涼。

5..... 將米連同香料放入攪拌機打碎，留意米要保持碎粒狀，避免打得過碎。

6..... 將米碎過篩，保留隔篩上較為粗粒的米碎。

7..... 把五花腩和米碎拌勻。

8..... 荷葉放於蒸籠上，番薯片放於底層，再排上五花腩，包好，大火蒸約 40-50 分鐘即成。

阿爺秘技 ———————————— Cooking tips

- 建議選用帶皮及偏瘦的五花腩，油脂太多的話，吃後會太膩口。
- 炒米時加入香料，令米碎帶香味，味道更多層次。
- 糯米及粘米勿打得過碎，甘香有味，有嚼勁。
- 五花腩醃至入味後，才加入米碎拌勻，吃時有不同的口感及層次。

- For this recipe, I prefer skin-on pork belly slightly on the leaner side. If it's too fatty, the steamed pork would taste too greasy after steamed.

- I add spices to the rice when stir-frying them. Not only do they add a lovely warmth and aroma to the rice, but also add another dimension to the flavour profile of this dish.

- Do not blend the rice too finely. Otherwise, you'd lose the unique aroma. You should still be able to bite into small grains of rice for mouthfeel after steamed.

- The ground rice mixture should be added to the pork after it has been marinated long enough. That would retain the different textures of the rice and the pork for layered effect.

阿爺廚房——消失的味道

Method

1. Soak the glutinous rice in water twice its volume for two hours. Soak dried lotus leaves in water till soft.
2. Slice pork belly thickly. Slice sweet potato.
3. Add marinade to the pork belly in the listed order. Mix well.
4. Drain the glutinous rice. Heat a dry wok. Stir-fry the glutinous rice and long-grain rice in the dry wok. Add star-anise, cassia bark and Sichuan peppercorns. Toss until the rice turns golden. Set aside to let cool.
5. Put the rice and the spices into a blender or food processor. Coarsely blend the mixture to retain some chunks of rice. You have over-done it if it turns into a powder.
6. Pass the ground rice mixture through wire mesh. Use only the chunky bits.
7. Add the ground rice to the marinated pork belly. Mix well.
8. Line a bamboo steamer with lotus leaves. Arrange the sliced sweet potato evenly over. Top with the pork belly. Wrap the lotus leaves into a package. Cover the lid of the steamer. Steam over high heat for 40 to 50 minutes.

網油腰肝拼臘腸卷

~Caul fat roll with kidney, liver and sausage~

菜式特點

「網油腰肝卷」於六、七十年代，曾是風行一時的菜式。當時的香港人生活水平較低，就用豬腰、雞肝等較平價的食材做菜；但由於這道菜工序繁複，加上現代人注重健康，所以越來越少人吃，這菜式就漸漸式微了。

【材料】

豬網油＿＿＿＿＿2 塊
豬腰＿＿＿＿＿＿90 克
雞肝＿＿＿＿＿＿90 克
加拿大臘腸＿＿＿30 克
金華火腿＿＿＿＿30 克
鮮冬菇＿＿＿＿＿4 朵（大）
唐芹＿＿＿＿＿＿10 克

【醃料】

玫瑰露酒＿＿＿＿1 湯匙
鹽＿＿＿＿＿＿＿1 茶匙
糖＿＿＿＿＿＿＿1 茶匙
生粉＿＿＿＿＿＿1 茶匙

Ingredients

- 2 sheets pork caul fat
- 90 g pork kidney
- 90 g chicken liver
- 30 g Canadian lean pork Cantonese-style sausage
- 30 g Jinhua ham
- 4 large fresh shiitake mushrooms
- 10 g Chinese celery

Marinade

- 1 tbsp Chinese rose wine
- 1 tsp salt
- 1 tsp sugar
- 1 tsp potato starch

【做法】

1..... 豬網油用水清洗數次，最後在水內放入少許玫瑰露酒浸泡一會，瀝乾水分，備用。

2..... 臘腸和金華火腿隔水蒸至軟身；唐芹切絲；鮮冬菇切片。

3..... 豬腰切半，去除中間白色及深紅色部分，剞花後切薄片，浸泡在加有白醋和玫瑰露酒的水中，去腥及異味。

4..... 切去雞肝的筋膜、脂肪，切厚片備用。

5..... 豬腰、雞肝用醃料醃 10 分鐘。

6..... 將蒸過的臘腸直切成四份，再切成適當長度；金華火腿切條，長度和臘腸相若。

7..... 豬網油鋪平，依次放入豬腰、雞肝、冬菇、金華火腿（或臘腸）、唐芹，捲成長條狀。

8..... 燒滾油，將網油腰肝卷在鑊邊輕放入鑊，以中細火慢炸，待轉成金黃色即可撈起，切件上碟。

Method

1. Rinse the caul fat in water a few times. Then soak it in water with a dash of Chinese rose wine for a while. Drain and set aside.
2. Steam pork sausage and Jinhua ham until soft. Set aside. Finely shred Chinese celery. Slice the shiitake mushrooms.
3. Cut the pork kidney in half lengthwise. Cut off the white membrane and all dark red parts inside. Make light crisscross incisions on the outside. Cut into thin slices. Soak in a bowl of water with white vinegar and Chinese rose wine in it to remove the unpleasant smell.
4. Remove the membranes on the chicken liver. Cut off the fat. Slice thickly.
5. Add marinade to the pork kidney and chicken liver. Mix well and leave them for 10 minutes.
6. Cut each steamed sausage into quarters lengthwise. Then cut into suitable lengths. Cut Jinhua ham into strips about the same length as the sausage.
7. Lay flat a sheet of caul fat. Put in pork kidney, chicken liver, shiitake mushrooms, Jinhua ham (or pork sausage) and Chinese celery in this particular order. Fold and roll the caul fat into a cylinder shape. Repeat with the other sheet of caul fat.
8. Heat oil in a wok. Slide the caul fat rolls gently into the oil. Deep-fry over medium-low heat until golden. Drain and slice. Serve.

阿爺秘技 ──────────── Cooking tips

- 豬網油較油膩，建議配搭瘦臘腸，以免吃後滿口油膩感。
- 必須徹底去掉豬腰白色及深紅色部分，而且豬腰剕斜紋更易熟透。
- 各材料的長度均等，有助掌握火候；如長短不一，受火程度不同，有些材料會因炸太久而變硬不好吃。
- 唐芹是這個餸點睛之料，令炸物加添清香之味。

- Pork caul fat is quite greasy. Thus, I use Canadian lean pork sausage to balance out the greasiness. If you use regular Cantonese pork sausage, the rolls may turn out too greasy.

- Make sure you cut off all of the white membrane and the dark red part inside the pork kidney. I also make crisscross cuts on it to make it cook faster.

- Try to cut every ingredient into roughly the same thickness and length. That would make it easier for you to control the deep-frying time. If deep-fried for too long, the filling ingredients would turn rubbery and too chewy.

- Chinese celery is the icing on the cake for this recipe – it makes deep-fried food less greasy and imparts a lovely fragrance.

春菜排骨煲

~Braised pork with Chun Cai in clay pot~

菜式特點

潮汕人只要到了立春就會用春菜入饌，因為它有迎春的意思。這道菜是潮州家庭常吃的菜式，多以肥豬肉配搭烹調，煮好後待翌日翻熱食用，味道特別濃郁、好味。

【材料】

腓排	600 克
春菜	750 克
燒腩肉	400 克
蒜蓉	40 克
薑片	60 克
鹽	2 茶匙
水	800 毫升

Ingredients

- 600 g premium boneless pork ribs
- 750 g Chun Cai (Swatow leaf mustard)
- 400 g roast pork belly
- 40 g grated garlic
- 60 g ginger (sliced)
- 2 tsp salt
- 800 ml water

阿爺秘技   Cooking tips

- 火腩帶油香，比燒肉骨位更適合炆煮春菜。
- 將燒腩肉爆香至肥肉部位呈透明，令油脂徹底釋出，才加入春菜。
- 宜挑選帶鮮綠色葉片的春菜，吸收油香肉味後更好味，味道清甜、甘香。
- 先加入春菜莖煮一會，待軟腍才下腓排及葉片續煮。

- Roast pork belly is the cut of choice for this recipe because of its high fat content. It complements Chun Cai way better than leaner roast pork.
- Fry the roast pork belly until its fat starts to turn transparent before adding Chun Cai. This step ensures enough pork fat is rendered in the wok for the Chun Cai to pick up.
- Pick Chun Cai leaves with bright green colour for this recipe. They pick up the pork fat nicely and would taste aromatic and sweet after braised.
- Put in the Chun Cai stems first and cook for a while till soft. Then add pork ribs and Chun Cai leaves and cook further.

春菜外形細長，葉子大且有紋理，煮後味道甘甜。
Chun Cai is long and slender in shape. Its leaves are large in size with wrinkly veins. It tastes sweet after cooked.

【做法】

1......燒腩肉、腓排切件；春菜切段，把根部拍扁。

2......腓排汆水約 10 分鐘，盛起。

3......起油鑊，爆香薑片、蒜蓉、燒腩肉，加入春菜莖炒至春菜轉色後，加入腓排、春菜葉、水及鹽，煮約 30 分鐘即成。

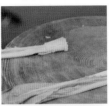

Method

1. Cut roast pork belly and pork ribs into pieces. Set aside. Cut Chun Cai into short lengths. Crush their roots with the flat side of a knife.

2. Blanch pork ribs in boiling water for about 10 minutes. Set aside.

3. Heat wok and add oil. Stir-fry ginger, garlic and roast pork belly until fragrant. Put in Chun Cai stems and stir until it turns colour. Add pork ribs, Chun Cai leaves, water and salt. Cook for 30 minutes. Save in a clay pot and serve.

蟬蛻牛腩煲

~Braised beef brisket with cicada shells~

菜式特點

蟬蛻含豐富蛋白質、氨基酸及甲殼素，加入牛腩裏，可以讓牛腩更鬆軟、更香，而且不會破壞牛腩的味道。

【材料】

崩沙腩（蝴蝶腩）＿＿＿＿＿900 克
蟬蛻＿＿＿＿＿＿＿＿＿＿10 隻
蓮藕＿＿＿＿＿＿＿＿＿＿600 克
炸枝竹＿＿＿＿＿＿＿＿＿120 克
冰糖＿＿＿＿＿＿＿＿＿＿100 克
柱侯醬＿＿＿＿＿＿＿＿＿2 湯匙
磨豉醬＿＿＿＿＿＿＿＿＿2 湯匙
薑＿＿＿＿＿＿＿＿＿＿＿5-6 片
乾葱＿＿＿＿＿＿＿＿＿＿3 粒
糖＿＿＿＿＿＿＿＿＿＿＿1.5 茶匙

Ingredients

- 900 g beef skirt brisket
- 10 dried cicada shells
- 600 g lotus root
- 120 g deep-fried beancurd sticks
- 100 g rock sugar
- 2 tbsp Chu Hau sauce
- 2 tbsp ground soybean paste
- 5 to 6 slices ginger
- 3 shallots
- 1.5 tsp sugar

阿爺秘技 ＿＿＿＿＿ **Cooking tips**

- 若全程用大火炆煮，材料會太硬太實，並且收乾水分，難以入口，宜用中細火慢煮。
- 炆約 1 小時後打開鍋蓋，有助散死氣及降溫，以免溫度太熱令牛腩變得硬實，肉質黏得太緊。
- The beef brisket has to be cooked for over 2 hours. If you cook it over high heat all the way through, the beef would turn rubbery and dry. Cuts with connective tissues should always be slow-cooked over medium-low heat.
- Open the lid once after braising the beef for 1 hour. That helps bring the temperature down and release any stale gas inside the pot. If the pot is too hot, the muscle fibres would tense up too much and the beef will be rubbery and dry.

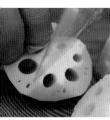

【做法】

1..... 崩沙腩切件，汆水，浸水略沖；蓮藕切件，略拍扁；蟬蛻放入茶袋內。

2..... 燒熱砂鍋，下油爆香薑片及乾葱，灑入糖，加入崩沙腩炒香，見肉邊位漸漸透明，下水至蓋過崩沙腩。

3..... 加入冰糖、蟬蛻，滾後煲 10-15 分鐘，下柱侯醬、磨豉醬拌勻，調至中火炆約 1 小時，加入蓮藕再炆約 1 小時。

4..... 打開鍋蓋，關火，再加蓋焗 15 分鐘，開火，翻滾後再焗 15 分鐘，重複此步驟 2-3 次，直至崩沙腩變腍。

5..... 最後加入炸枝竹，炆約 20 分鐘即成。

Method

1. Cut the beef brisket into chunks. Blanch in boiling water. Drain and soak in cold water. Rinse well. Set aside. Cut lotus root into chunks. Crush gently with the flat side of a knife. Set aside. Put the dried cicada shells into a muslin bag. Tie well.

2. Heat a clay pot. Add oil and stir-fry ginger and shallot until fragrant. Add sugar and beef briskets. Toss until fragrant and the tendons turn transparent slightly. Add enough water to cover the beef.

3. Add rock sugar and dried cicada shells. Bring to the boil and keep cooking for 10 to 15 minutes. Add Chu Hau sauce and ground soybean paste. Mix well and turn to medium heat. Braise for about 1 hour. Add lotus root and simmer for another 1 hour.

4. Open the lid once. Turn off the heat. Cover the lid and leave it for 15 minutes. Turn on the heat again and bring to the boil. Leave it with the lid covered for 15 minutes. Repeat this step for 2 to 3 times until the beef brisket is tender.

5. Put in deep-fried beancurd sticks. Simmer for 20 minutes. Serve the whole pot.

阿爺粟米肉粒飯

~Grandpa's steamed rice with pork cubes and sweet corn~

菜式特點

這是老爸年代的粟米肉粒飯版本，各材料顆粒分明、乾身，現在已很少人炮製這道菜了。

【材料】

新鮮粟米_____100 克
急凍豬柳_____150 克
五香豆乾_____30 克
菜脯_____30 克
甘筍_____30 克
青甜椒_____30 克
松子仁_____30 克
白飯_____1 碗

【醃料】

糖_____半茶匙
鹽_____半茶匙
胡椒粉_____1 茶匙
生抽_____1 湯匙
紹興酒_____1 茶匙

Ingredients

- 100 g fresh sweet corn
- 150 g frozen pork loin
- 30 g five-spice dried tofu
- 30 g dried radish
- 30 g carrot
- 30 g green bell pepper
- 30 g pine nuts
- 1 bowl steamed rice

Marinade

- 1/2 tsp sugar
- 1/2 tsp salt
- 1 tsp ground white pepper
- 1 tbsp light soy sauce
- 1 tsp Shaoxing wine

阿爺秘技 ——————————— Cooking tips

- 此菜式重點在於材料配搭，咬入口，嚐到五種不同的口感。
- 豬柳要切稍大一點，炒熟後收縮至與粟米相若的大小。
- 不要加入芡汁，否則炒材料時黏着不能推開，做不成粒粒分明的效果。

- The key of this recipe lies in the combination of ingredients. You should be able to taste all five different mouthfeels in every bite.
- You may dice the pork slightly more coarsely than the sweet corn kernels. The pork tends to shrink after cooked and the pork cubes will be of the same size as sweet corn kernels when served.
- Do not finish up with a thickened glaze. Otherwise, the ingredients would clump together and you don't get to enjoy the crisp texture of every dice of ingredients.

【做法】

1..... 急凍豬柳解凍、切粒,下醃料醃 30 分鐘。
2..... 白鑊下松子仁烘乾至金黃色,備用。
3..... 新鮮粟米、豆乾、菜脯、甘筍、青甜椒全部
切粒。
4..... 燒熱油鑊,放入粟米、甘筍、豆乾、菜脯及
豬柳粒炒至七分熟,加入青椒粒炒勻。
5..... 上碟後,灑入松子仁,配白飯供吃。

Method

1. Thaw the pork loin. Dice it. Add marinade and mix well. Leave it for 30 minutes.
2. Toast the pine nuts in a dry wok until golden. Set aside.
3. Cut the kernels off the cobs of the sweet corn. Dice dried tofu, dried radish, carrot and green bell pepper.
4. Heat wok and add oil. Stir-fry sweet corn kernels, carrot, dried tofu, dried radish and pork loin until medium-well done. Put in green bell pepper. Toss again.
5. Save on a serving plate. Sprinkle with pine nuts. Serve with steamed rice on the side.

水晶豬手凍

~Pork trotter aspic~

菜式特點

「水晶豬手凍」是江蘇鎮江的傳統名菜，流傳於江淮一帶，至今已有差不多 300 年歷史。以前會用硝醃製豬手，令肉質呈現紅色，讓豬手看起來優質新鮮，但由於這種物質對人體有害，現已不再使用。

【材料】

豬手	1 隻
豬皮	1 塊
花椒	1 湯匙
八角	4 粒
桂皮	2 塊
香葉	2 片
草果	4 粒
陳皮	2 角
白胡椒粒	1 湯匙
冰糖	30 克
葱	2-3 棵
薑	3-4 片

Ingredients

- 1 pork trotter
- 1 sheet fresh pork skin
- 1 tbsp Sichuan peppercorns
- 4 pods star-anise
- 2 pieces cassia bark
- 2 bay leaves
- 4 Tsaoko fruits
- 2 pieces aged tangerine peel
- 1 tbsp white peppercorns
- 30 g rock sugar
- 2 to 3 sprigs spring onion
- 3 to 4 slices ginger

【做法】

1..... 預備一鍋薑蔥水，放入豬皮煮約 1-1.5 小時，豬皮軟至可用筷子戳穿。

2..... 豬皮切走脂肪後切條；陳皮浸軟，備用。

3..... 豬皮用粗鹽加清水搓洗，再過水洗淨，重複搓洗至水清為止。

4..... 預備蒸鑊，放入一大碗清水燉熱，下糖及鹽，加入豬皮至水剛蓋過豬皮面，燉約 2 小時，讓豬皮釋出膠質，煮成豬皮膠汁。

5..... 預備一鍋沸水，加入桂皮、八角、花椒、白胡椒粒、草果、香葉、陳皮，煮至出味，加入冰糖，滷煮豬手約 2 小時至腍身。

6..... 豬手拆肉後放涼。

7..... 預備長方形盆，放入豬手肉後，倒入（4），用保鮮紙包好，冷藏至凝固，取出切件即成。

阿爺秘技 ———————————— Cooking tips

- 豬皮用粗鹽加清水搓洗，換水，再洗至沒有雜質浮出才拿去燉煮。經燉煮熬出的豬皮膠汁可以令成品晶瑩剔透（圖 1）。其實，豬皮除了可燉煮取膠汁外，還可以用煲湯的方法取膠汁，但成品會比較混濁（圖 2）。

- To keep the aspic crystal clear, make sure you rub the pork skin repeatedly with coarse salt and rinse the pork skin with water until water runs clear, before double-boiling it in the water bath (picture 1). Instead of double-boiling it, you may also cook the pork skin in boiling water heated directly over the stove. But the aspic tends to look more cloudy and milky in colour this way (picture 2).

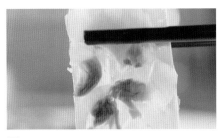 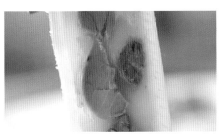

圖 1 (picture 1)　　　　　　　　圖 2 (picture 2)

Method

1. Boil a pot of water. Add spring onion and ginger. Put in the pork skin. Cook for 1 to 1.5 hours over medium-low heat until you can pierce through the pork skin with a chopstick.
2. Trim off any fat on the pork skin. Cut into strips. Soak aged tangerine peel in water until soft.
3. Add coarse salt to pork skin. Add water, rub and scrub well. Drain and rinse with fresh water. Rub salt, scrub and rinse repeatedly until the water runs clear.
4. Boil water in a wok. Put a bowl of water into the simmering water. Add sugar and salt to the water in the bowl. Put the pork skin strips into the bowl. The water in the bowl should be enough to cover all pork skin. Cover the lid and double-boil the pork skin for 2 hours to extract the gelatine.
5. Boil a pot of water. Put in cassia bark, star anise, Sichuan peppercorns, white peppercorns, Tsaoko fruits, bay leaves and aged tangerine peel. Cook until the flavours are infused. Add rock sugar and put in the pork trotter. Bring to the boil and cook for 2 hours over low heat until tender.
6. De-bone the pork trotter. Break the meat into small pieces with a fork. Leave the meat to cool.
7. Prepare a rectangular tray. Arrange the pork trotter meat on the bottom of the tray evenly. Pour the gelatinous pork skin soup from step 4 over. Leave it to cool. Cover in cling film and refrigerate until set. Slice and serve.

▶ 示範短片

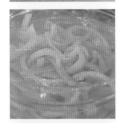

腐乳炸排骨

~Fermented beancurd-flavoured deep-fried pork ribs~

菜式特點

據史料記載，早在公元 5 世紀魏代古籍中，有腐乳生產
工藝的記載，到了明代就大量加工腐乳。腐乳的蛋白質
和鈣含量很高，是一種很有營養的發酵食品。

【材料】

原味腐乳_____8 磚
一字排_____500 克
糖_____3 茶匙
米酒_____3 湯匙
麵粉_____適量

Ingredients

- 8 cubes fermented beancurd
- 500 g pork spareribs
- 3 tsp sugar
- 3 tbsp rice wine
- plain flour

【醃料】

糖_____2 茶匙
鹽_____1 茶匙
紹興酒_____半茶匙
胡椒粉_____少許
生粉_____半茶匙
油_____2 茶匙

Marinade

- 2 tsp sugar
- 1 tsp salt
- 1/2 tsp Shaoxing wine
- ground white pepper
- 1/2 tsp potato starch
- 2 tsp oil

原味腐乳
Fermented beancurd

這是腐乳發酵的重要過程。
This is the important fermentation process of fermented beancurd.

辣椒腐乳
Fermented beancurd with chillies

【做法】

1..... 一字排切件，順次序下糖、鹽、紹興酒、胡椒粉、生粉、油拌勻。

2..... 腐乳加入米酒、糖拌勻成糊狀。

3..... 一字排沾上麵粉，再上腐乳糊，最後再上麵粉。

4..... 起油鑊，下一字排炸至金黃，完成。

Method

1. Chop pork spareribs into chunks. Add the marinade ingredients in the order listed. Mix well.

2. Add rice wine and sugar to the fermented beancurd. Mash and stir into a paste.

3. Coat the pork ribs in flour. Then coat them in the fermented beancurd paste. Coat them in flour once more.

4. Heat oil in a wok. Deep-fry the pork ribs until golden. Serve.

阿爺秘技 ——————— Cooking tips

- 選用一字排的原因是，它只有三成左右肥肉，其餘是瘦肉，拿來蒸也好、炸也好，肉質軟腍。
- 炸排骨前先上乾粉，有助吸附腐乳醬。
- 醃排骨時加入適量油，排骨炸後肉質更鬆軟。

■ I use spareribs because the fat to lean meat ratio is about 3 to 7. It is tender and delicious no matter deep-fried or steamed.

■ Coat the ribs in flour first so that fermented beancurd paste can cling on to them.

■ I add some oil to the marinade, as the oil keep the meat juicy and moist after deep-fried.

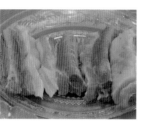  

脆皮牛肋條

~Crispy fried beef rib fingers~

菜式特點

自製滷水汁炆煮牛肋條,利用餘溫的熱力浸焗,令牛肋條更加軟腍入味,最後蘸上粉漿炸至香脆,演變出創新的吃法。

【材料】

牛肋條＿＿＿＿＿＿500 克
薑＿＿＿＿＿＿＿＿4-5 片
蒜頭＿＿＿＿＿＿＿6-7 粒
乾葱＿＿＿＿＿＿＿4 粒
冰糖＿＿＿＿＿＿＿30 克
紹興酒＿＿＿＿＿＿100 毫升
魚露＿＿＿＿＿＿＿1 湯匙
生抽＿＿＿＿＿＿＿適量
老抽＿＿＿＿＿＿＿1-2 湯匙

Ingredients

- 500 g beef rib fingers
- 4 to 5 slices ginger
- 6 to 7 cloves garlic
- 4 shallots
- 30 g rock sugar
- 100 ml Shaoxing wine
- 1 tbsp fish sauce
- light soy sauce
- 1 to 2 tbsp dark soy sauce

【滷水料】

八角＿＿＿＿＿＿＿2 粒
香葉＿＿＿＿＿＿＿2 片
桂皮＿＿＿＿＿＿＿4 克
草果＿＿＿＿＿＿＿4 粒
甘草＿＿＿＿＿＿＿4 片
白胡椒粒＿＿＿＿＿3 克
花椒粒＿＿＿＿＿＿5-10 克
丁香＿＿＿＿＿＿＿3 克

Marinade ingredients

- 2 pods star-anise
- 2 bay leaves
- 4 g cassia bark
- 4 Tsaoko fruits
- 4 slices liquorice
- 3 g white peppercorns
- 5 to 10 g Sichuan peppercorns
- 3 g cloves

【炸漿料】

生粉＿＿＿＿＿＿＿100 克
粟粉＿＿＿＿＿＿＿100 克
發粉＿＿＿＿＿＿＿5 克
蛋黃＿＿＿＿＿＿＿1 個
水＿＿＿＿＿＿＿＿100 克

Deep-frying batter

- 100 g potato starch
- 100 g cornstarch
- 5 g baking powder
- 1 egg yolk
- 100 g water

【做法】

1..... 牛肋條汆水後洗淨,瀝乾水分。

2..... 另起油鑊,爆香乾葱、薑、蒜頭、花椒、桂皮、香葉、白胡椒粒、丁香、八角、甘草、草果,灒入紹興酒,加入冰糖,適量沸水,待冰糖溶化後,下生抽及老抽調色成滷水。

3..... 把牛肋條加入滷水中,下魚露,翻滾後轉慢火,煮 35-40 分鐘,關火,焗 6-7 小時至入味。

4..... 用生粉、粟粉、發粉、蛋黃、水拌勻成炸漿。

5..... 牛肋條瀝乾,沾上生粉。

6..... 準備油鑊,等待適合油溫。

7..... 牛肋條上炸漿,落鑊炸至金黃色,撈起。

8..... 待稍為降溫,切件,灑上椒鹽,伴以唸汁食用。

Method

1. Blanch the beef rib fingers. Rinse well and drain.
2. Heat a wok and add oil. Stir-fry shallots, ginger, garlic, Sichuan peppercorns, cassia bark, bay leaves, white peppercorns, cloves, star-anise, liquorice and Tsaoko fruits. Add Shaoxing wine and rock sugar. Pour in some boiling water. Cook until sugar dissolves. Add light soy sauce and dark soy sauce. Mix well to make the spiced marinade.
3. Put the beef rib fingers into the spiced marinade. Add fish sauce and bring to the boil. Turn to low heat. Simmer for 35 to 40 minutes. Turn off the heat and leave the beef rib fingers in the marinade for 6 to 7 hours until flavourful.
4. In a bowl, mix potato starch, cornstarch, baking powder, egg yolk and water into deep-frying batter.
5. Remove the beef rib fingers from the spiced marinade. Wipe dry. Coat them in potato starch.
6. Heat a wok of oil until hot enough for deep-frying.
7. Dunk the beef rib fingers into the deep-frying batter from step 4. Deep-fry in hot oil until golden. Drain and set aside.
8. Leave the beef rib fingers to cool slightly. Slice and sprinkle with peppered salt. Serve with Worcestershire sauce on the side.

阿爺秘技 _____ **Cooking tips**

- 用滷水汁先將牛肋條煮熟，用餘溫烹煮讓它入味，這樣牛肋肉不會因炆煮過久而變韌及散開。
- I cook the beef rib fingers in spiced marinade first, and then let them steep in the marinade to pick up the flavours. This way, the beef rib fingers would be flavourful, but would not be overcooked and turn rubbery.

糖醋五花腩

~Braised pork belly in sweet and sour sauce~

菜式特點

很多人都害怕五花腩的脂肪，不過用了糖醋就可以令它
變得更軟，而且把油膩逼出來。

【材料】

五花腩_____800 克	八角_____1 粒
片糖_____約半片 -1 片	冰糖_____40 克（煮滷水用）
鎮江醋_____2 湯匙	紹興酒_____3 湯匙
甘筍_____1 條	鹽_____1 茶匙
西芹_____2 條	老抽_____1 湯匙
香葉_____1 片	冰糖_____40 克（煮焦糖用）
桂皮_____1 塊	

Ingredients

- 800 g pork belly
- 1/2 to 1 raw cane sugar slab
- 2 tbsp Zhenjiang vinegar
- 1 carrot
- 2 celery stems
- 1 bay leaf
- 1 piece cassia bark

- 1 pod star-anise
- 40 g rock sugar (for spiced marinade)
- 3 tbsp Shaoxing wine
- 1 tsp salt
- 1 tbsp dark soy sauce
- 40 g rock sugar (for caramel used to colour the pork)

【做法】

1..... 五花腩汆水，洗淨，瀝乾水分。

2..... 甘筍切件、西芹切段。

3..... 預備 3-3.5 公升沸水，加入甘筍、西芹、香葉、桂皮、八角煮至出味成滷水，加入冰糖、紹興酒、鹽、老抽拌勻，煮成蔬菜滷汁備用。

4..... 鑊中下油，加入冰糖煮成焦糖，下五花腩上色。

5..... 將已上色的五花腩加入滷水中，煮約 1.5 小時，取出切成正方形。

6..... 於鑊中加入 500 毫升滷水汁，加入鎮江醋、半片至 1 片的片糖，視乎喜歡的酸甜度，煮勻成糖醋汁，下五花腩煮至收汁，完成。

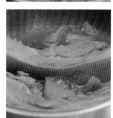

Method

1. Blanch the pork in boiling water. Rinse well and drain.
2. Cut carrot into chunks. Cut celery into short lengths.
3. Boil 3 to 3.5 litres of water. Put in carrot, celery, bay leaf, cassia bark and star-anise. Cook until the flavours are infused. Add rock sugar, Shaoxing wine, salt and dark soy sauce. Mix well. This is the spiced marinade.
4. Heat a wok and add oil. Put in the rock sugar and cook until caramelized. Put in the pork belly to colour evenly on all sides.
5. Put the pork into the spiced marinade from step 3. Cook over low heat for 1.5 hours. Cut the pork into a neat square.
6. Transfer 500 ml of spiced marinade into a wok. Add Zhenjiang vinegar and raw cane sugar slab. Bring to the boil and cook until sugar dissolves. Taste it and season according to your preferred sweetness or sourness. Put in the pork belly. Cook till the sauce is reduced. Serve.

 **Cooking tips**

- 煮上色料前先把冰糖拍碎，用慢火煮至溶化，放入五花腩煎一會，就會呈現漂亮的金黃色。
- 醋和酒可減低五花腩的油膩感，吃時更感清爽。
- To colour the pork belly with caramel, crush the rock sugar first. Cook over low heat until caramelized. Then put the pork in to fry on each side briefly. The pork would then turn beautifully golden brown.
- Adding vinegar and wine helps cut through the richness of the pork belly. It would tastes less greasy that way.

紙包雞

~Paper-bag chicken~

菜式特點

「紙包雞」起源於清朝，在梧州民間始創。自清朝咸豐年間起，被列為梧州府府台宴客的菜式。這次採用本地飼養的鮮雞炮製「紙包雞」，斬件入紙袋炸，味道濃郁，外脆內嫩。

【材料】

鮮雞＿＿＿＿＿＿半隻
葱＿＿＿＿＿＿＿50 克
乾葱＿＿＿＿＿＿50 克

Ingredients

- 1/2 freshly slaughtered chicken
- 50 g spring onion
- 50 g shallots

【醃料】

薑＿＿＿＿＿＿＿50 克（磨汁）
米酒　＿＿＿＿＿2 湯匙
沙薑粉＿＿＿＿＿1 湯匙
胡椒粉＿＿＿＿＿1 茶匙
五香粉＿＿＿＿＿1 湯匙
蠔油＿＿＿＿＿＿2 湯匙
麻油＿＿＿＿＿＿少許
生粉＿＿＿＿＿＿2 湯匙

Marinade

- 50 g ginger (grated and juice squeezed)
- 2 tbsp rice wine
- 1 tbsp sand ginger powder
- 1 tsp ground white pepper
- 1 tbsp five-spice powder
- 2 tbsp oyster sauce
- sesame oil
- 2 tbsp potato starch

【工具】

牛油紙袋＿＿＿＿數個

Utensils

- a few parchment cooking bags (or made with baking paper)

【做法】

1..... 鮮雞洗淨、斬件，瀝乾水分，加入薑汁、米酒拌勻，依次下沙薑粉、胡椒粉、五香粉、蠔油及麻油，完全拌勻後，加入生粉拌和，最後下少許水讓雞件吸收，放入雪櫃醃 2 小時。

2..... 葱切度；乾葱剁成蓉。

3..... 將醃好的雞件與乾葱蓉拌勻，放進預先準備的牛油紙袋，加少許葱度，用竹籤封口備用。

4..... 鍋中下油，用中火燒滾，放入紙包雞炸至金黃即成。

Method

1. Rinse the chicken and chop into pieces. Drain well. Add ginger juice and rice wine. Mix well. Then add sand ginger powder, ground white pepper, five-spice powder, oyster sauce and sesame oil in this particular order. Mix well. Then add potato starch and mix again. Add a dash of water at last and let the chicken pick up the moisture. Refrigerate for 2 hours.
2. Cut spring onion into short lengths. Finely chop shallots.
3. Add shallots to the chicken. Mix well. Transfer into parchment cooking bags. Top with a couple sprigs of spring onion in each bag. Seal with bamboo skewers. Set aside.
4. Heat oil in a wok. Put on medium heat until hot enough. Deep-fry the chicken in paper bags until the chicken turns golden. Serve.

阿 爺 秘 技 ————— **Cooking tips**

- 待醃料拌勻雞件後，才放入生粉，以免變成糊狀。
- 乾葱的味道太霸道，故雞件冷藏後才最後加入輕拌。
- 雞件放入牛油紙袋，令雞肉皮脆內嫩，更有油炸的香味。
- 炸紙包雞不宜用猛火，以免外焦內生，緊記使用中火烹調。
- When you marinate the chicken, stir in the rest of the marinade ingredients first before adding potato starch. Otherwise, the starch would pick up the seasoning and turn into a paste.
- Shallots have very strong aromas. Thus, it's not advisable to marinate the chicken with shallots for a long period of time. That's why I stir in the shallots only after the chicken has been refrigerated for 2 hours.
- Deep-frying the chicken inside parchment bags keeps the chicken tender and juicy, while still retain a crispy crust.
- Do not deep-fry the chicken in paper bags over high heat. Otherwise, the chicken may burn on the outside while the inside is still raw. Make sure you use medium heat.

豆豉雞煲

~Chicken with fermented black beans in clay pot~

菜式特點

這次炮製的「豆豉雞煲」，首先爆香料頭及鮮雞，全個過程不加入水焗煮，保留雞的精華，最後加蓋淋上紹興酒，上枱前香氣四溢，滋味濃郁！

【材料】

鮮雞＿＿＿＿＿＿＿1 隻

豆豉＿＿＿＿＿＿＿25 克

薑片＿＿＿＿＿＿＿50 克

乾葱＿＿＿＿＿＿＿4 粒

蒜頭＿＿＿＿＿＿＿4 粒

葱＿＿＿＿＿＿＿5 棵

芫茜＿＿＿＿＿＿＿5 棵

指天椒＿＿＿＿＿＿2 隻

乾辣椒＿＿＿＿＿＿2 隻

Ingredients

- 1 freshly slaughtered chicken
- 25 g fermented black beans
- 50 g sliced ginger
- 4 shallots
- 4 cloves garlic
- 5 sprigs spring onion
- 5 sprigs coriander
- 2 bird's eye chillies
- 2 dried chillies

【調味料】

糖＿＿＿＿＿＿＿1 茶匙

鹽＿＿＿＿＿＿＿1 茶匙

紹興酒＿＿＿＿＿＿2 湯匙

蠔油＿＿＿＿＿＿＿1 湯匙

老抽＿＿＿＿＿＿＿1 湯匙

Seasoning

- 1 tsp sugar
- 1 tsp salt
- 2 tbsp Shaoxing wine
- 1 tbsp oyster sauce
- 1 tbsp dark soy sauce

阿爺秘技 —————— Cooking tips

- 豆豉經過油炸，可保持原粒狀，不會與雞件糊成一團。
- 整個烹調過程不用加水，因加入糖及鹽調味後，會釋出水分，最後煮至水分略收乾。

- I deep-fry the fermented black beans to retain their shapes, so that they won't turn mushy when cooked with the chicken.
- You don't need to add water to this dish throughout the whole cooking process. After you season the chicken with sugar and salt, moisture would be drawn out of the meat. Just cook till the moisture almost dries out.

【做法】

1..... 乾葱、蒜頭、葱切成料頭,備用。
2..... 雞洗淨,斬件,抹乾水分。
3..... 瓦鍋中加油,油滾後加入豆豉炸香,盛起備用。
4..... 瓦鍋內加入豆豉炸油,爆香薑片、乾葱及蒜頭,
      下雞件煎至兩面金黃(約四成熟)。
5..... 下糖、鹽、乾辣椒、指天椒、紹興酒、蠔油、豆
      豉及老抽炒勻,加蓋,煮至水分收乾及入味。
6..... 加入葱段,加蓋,於蓋面澆上紹酒,待 1 分鐘,
      最後加入芫荽即成。

Method

1. Finely chop shallots, garlic and spring onion. Set aside.
2. Rinse the chicken. Chop into pieces. Wipe dry.
3. Heat a clay pot and add oil. When the oil is smoking hot, deep-fry the fermented black beans until fragrant. Set aside the fermented black beans.
4. In the same clay pot, heat the oil used to deep-fry fermented black beans. Stir-fry ginger, shallot and garlic. Fry the chicken until both sides golden and about medium-rare cooked.

5. Add sugar, salt, dried chillies, bird's eye chillies, Shaoxing wine, oyster sauce, fermented black beans and dark soy sauce. Toss well and cover the lid. Cook until the liquid reduces and the chicken picks up the seasoning.

6. Add spring onion and cover the lid. Drizzle with Shaoxing wine over the lid. Wait for 1 minute. Garnish with coriander at last. Serve the whole pot.

芋蓉鴨

~Deep-fried boneless duck stuffed with mashed taro~

菜式特點

「芋蓉鴨」源自於潮州，正名為「芋茸酥鴨」，於五、六十年代的香港非常流行。芋蓉加入了澄麵及其他配料，炸起來如蜂巢一樣，酥脆香口。

【材料】

草鴨＿＿＿＿＿＿1 隻
芋頭＿＿＿＿＿＿600 克
澄麵＿＿＿＿＿＿185 克
豬油＿＿＿＿＿＿185 克
食用臭粉＿＿＿＿3 克
梳打粉＿＿＿＿＿3 克
鹹蛋黃＿＿＿＿＿2-3 個（蒸熟）
五香粉＿＿＿＿＿10 克
大地魚粉＿＿＿＿10 克
蝦米＿＿＿＿＿＿10 克
八角＿＿＿＿＿＿3 粒
桂皮＿＿＿＿＿＿5 克
香葉＿＿＿＿＿＿4 片
薑＿＿＿＿＿＿＿7-8 片
陳皮＿＿＿＿＿＿1 個

【調味料】

老抽＿＿＿＿＿＿60 毫升
糖＿＿＿＿＿＿＿2 茶匙
鹽＿＿＿＿＿＿＿1 茶匙

Ingredients

- 1 "grass duck"
- 600 g taro
- 185 g wheat starch
- 185 g lard
- 3 g baker's ammonia
- 3 g baking soda
- 2 to 3 salted egg yolks (steamed till done)
- 10 g five-spice powder
- 10 g ground dried plaice
- 10 g dried shrimps
- 3 pods star-anise
- 5 g cassia bark
- 4 bay leaves
- 7 to 8 slices ginger
- 1 whole aged tangerine peel

Seasoning

- 60 ml dark soy sauce
- 2 tsp sugar
- 1 tsp salt

【做法】

1..... 草鴨汆水，略放涼，在鴨身塗抹老抽上色，待風乾。

2..... 芋頭切去兩端，取中間部分，去皮、切片，蒸 20-30 分鐘。

3..... 蝦米放入攪拌器打成粉狀。陳皮浸軟，備用。

4..... 起油鑊，將熱油淋於鴨身至金黃色，待涼。

5..... 鴨肚內放入陳皮、香葉、桂皮、八角、薑片；鴨身均勻塗抹五香粉，放於蒸碟上，於鴨面鋪上陳皮及薑片，蒸約 1.5-2 小時至腍身。

6..... 取出全鴨，去骨，盡量保留一塊完整鴨肉（連鴨皮），放在碟內。

7..... 芋頭壓成芋泥，加入豬油、梳打粉、臭粉搓勻，再下大地魚粉、蝦米粉、糖、鹽、鹹蛋黃搓勻。

8..... 澄麵用熱水拌勻成熟澄麵糰，加入芋泥內搓勻。

9..... 將芋泥鋪釀在已拆骨的連皮鴨肉上。

10... 起油鑊，將熱油淋於鴨上，直至底面呈金黃及蜂巢狀，隔油約 20 分鐘，切件上碟即成。

Method

1. Blanch the duck in boiling water. Leave it to cool briefly. Brush dark soy sauce all over the skin. Hang to air-dry.
2. Cut both ends off the taro and use only centre. Peel and slice it. Steam for 20 to 30 minutes.
3. Blend the dried shrimps in a blender until powder-like. Soak aged tangerine peel in water until soft.
4. Heat a wok and add oil. Put the duck on a strainer ladle. Ladle hot oil and pour on the duck skin until the duck is nicely browned. Set aside to let cool.
5. Stuff the duck with aged tangerine peel, bay leaves, cassia bark, star-anise and ginger. Rub five-spice powder evenly on the skin. Transfer the duck on a steaming plate. Arrange aged tangerine peel and sliced ginger on top. Steam for 1.5 to 2 hours until tender.
6. Let the duck cool briefly. De-bone it while try to keep the duck meat in one piece without breaking the skin. Save on a plate with the skin side down.
7. Mash the steamed taro. Add lard, baking powder and baker's ammonia. Knead to mix well. Add ground dried plaice, dried shrimps, sugar, salt and mashed salted egg yolks. Knead to mix well.
8. Add hot water to wheat starch. Mix well into dough. Add to the mashed taro. Knead to incorporate.
9. Spread the mashed taro mixture over the de-boned duck over the meat side.
10. Heat wok and add oil. Put the stuffed duck on a strainer ladle with the skin side down. Ladle hot oil and pour over the stuffed duck and dip the skin side into the hot oil from time to time until both sides golden and puffs up with crumbly crust. Set aside the stuffed duck on the strainer ladle to drain the oil for 20 minutes. Slice into pieces. Serve.

- 草鴨的肉質結實,適合製作這道菜;如使用普通的鴨,則容易散開來。
- 600克芋頭需要拌入各185克澄麵及豬油,才有蜂巢的效果。
- 芋頭兩端切掉不要,取中間部分較粉糯好吃。
- 澄麵用熱水搓成熟麵糰,再加入芋泥內搓勻,緊記要搓得剛好黏起來,不要太稀。

- "Grass duck" has firmer meat, and is perfect for this recipe. If you use regular duck, the meat may break apart in the cooking process.
- To achieve the crumbly crust on the mashed taro, add 185 g of wheat starch and 185 g of lard to every 600 g of taro.
- Cut off and discard both ends of a taro as they are crunchy in texture. Only the centre part is starchy and works for this recipe.
- You have to scald the wheat starch in hot water and knead into dough before adding it to the mashed taro. Make sure the mashed taro has a thick and sticky consistency without being too runny.

陳皮檸檬燉鴨

~Double-steamed duck soup with aged tangerine peel and salted lemon~

菜式特點

草鴨配搭陳皮及鹹檸檬燉成湯，令湯水帶有濃濃的陳皮香氣，具有祛痰順氣的功效。

【材料】

草鴨_____1 隻
陳皮_____5 角
鹹檸檬_____2 個
青檸檬_____1 個

Ingredients

- 1 "grass duck"
- 5 pieces aged tangerine peel
- 2 salted lemons
- 1 lime

阿爺秘技 — Cooking tips

- 用熱開水燉湯，香氣更快散發出來。
- 鴨件必須汆水，燉出來的湯才清澈。
- I add boiling hot water to make double-steamed soups, so that the flavours can be infused sooner.
- Make sure you blanch the duck before using it to make double-steamed soup. Otherwise, the soup would turn out murky instead of clear.

【做法】

1..... 鴨洗淨，去掉尾部、鴨腳及龍骨，斬成四件，汆水，洗淨備用。
2..... 陳皮浸軟，備用。
3..... 準備燉盅，放入 1 個鹹檸檬、3 角陳皮、鴨件，於鴨件上再放上鹹檸檬 1 個（連汁）及陳皮 2 角。
4..... 倒入約 1 公升熱開水，用耐熱保鮮紙封好，加蓋，燉約 4.5 小時即成。享用時可灑入青檸檬汁。

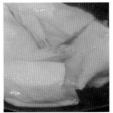

Method

1. Rinse the duck. Cut off the tail, duck feet and the spine. Cut into quarters. Blanch in boiling water. Drain and rinse well.

2. Soak aged tangerine peel in water till soft. Set aside.

3. In a double-steaming pot, put 1 salted lemon, 3 pieces of aged tangerine peel and the duck. Then arrange 1 salted lemon and its juices, with 2 pieces of aged tangerine peel on top.

4. Pour in 1 litre of boiling hot water. Seal with microwave-safe cling film. Cover the lid and double-steam in a pot of boiling water for 4.5 hours. Add the lime juice when serving.

第二章

海鮮、河鮮

這章節介紹的海鮮、河鮮菜式如膏蟹蒸肉餅、廣東蔗蝦、黃鱔焗飯、番茄蜆湯、海膽蒸蛋白等等，是鼎爺的心頭好。菜式的味道鮮甜，做法簡易，跟着鼎爺的詳細做法，想煮出美味的菜式，絕對難不倒你。

膏蟹蒸肉餅

~Steamed mud crab on ground pork patty~

菜式特點

將蟹膏、蟹肉與肉餅拌勻，再在上面放上膏蟹一起蒸，讓蟹汁和蟹膏的香味滲進去，成為升級版的「膏蟹蒸肉餅」。

【材料】

膏蟹＿＿＿＿＿＿2 隻
豬柳脢＿＿＿＿＿370 克

【醃料】

糖＿＿＿＿＿＿＿1 茶匙
鹽＿＿＿＿＿＿＿1 茶匙
油＿＿＿＿＿＿＿1 湯匙
水＿＿＿＿＿＿＿5 湯匙
生粉＿＿＿＿＿＿1 茶匙
胡椒粉＿＿＿＿＿少許

Ingredients

- 2 female mud crabs
- 370 g pork tenderloin

Marinade

- 1 tsp sugar
- 1 tsp salt
- 1 tbsp oil
- 5 tbsp water
- 1 tsp potato starch
- ground white pepper

【做法】

1..... 將一隻蟹起蓋，取出蟹膏，蟹身斬件，把取出的蟹膏蒸約 10 分鐘至熟透。
2..... 另一隻蟹斬件，蒸熟後拆肉。
3..... 豬柳胸一半剁蓉，一半剁成碎粒。
4..... 豬肉拌勻，依次序加入醃料拌勻，下蟹膏及蟹肉，再加入水拌勻。
5..... 將肉餅放於碟上，再放上生蟹件拼成蟹形，蒸約 20 分鐘即成。

Method

1. Remove the carapace of one crab. Scoop out the roe. Remove the gills and chop the body into pieces. Steam only the roe for 10 minutes until done.
2. Dress the other crab. Chop into pieces. Steam till done. Pick out the crabmeat and discard the shells.
3. Finely chop half of the pork. Then coarsely dice the remaining half.
4. Mix the pork well. Then add the marinade ingredients in the order listed. Mix well. Add the steamed crab roe from step 1 and the crabmeat from step 2. Add water and mix well.
5. Arrange the pork mixture on a steaming plate. Shape into a round patty. Re-assemble the raw crab pieces from step 1 over the pork patty to look like a whole crab. Steam for 20 minutes. Serve.

阿爺秘技 — Cooking tips

- 把蟹肉混合在肉餅中,味道更鮮香。
- 將豬柳脢切粒及剁蓉,拌勻後的肉餅更有口感。
- 蒸肉餅多加水分,令蒸後的肉餅鬆軟、滑溜。
- 肉餅鋪上碟後,在肉面戳數個洞有疏氣之效果,令肉餅受熱更均勻。

- Mixing the crabmeat in the pork patty gives it an extra dimension of umami and aroma.
- I finely chop half of the pork and coarsely dice the rest. That would give the pork patty a more complex texture.
- Make sure you add more water to the pork mixture before steaming. That's the key to fluffy and velvety pork patty after steamed.
- After you put the pork mixture on a steaming plate, make a few indentations on the surface with a chopstick. That would create air pockets for steam to circulate through. The pork patty would pick up the heat more evenly that way.

濃蝦湯乾炒伊麵

~Stir-fried E-fu noodles in lobster broth~

菜式特點

現在極少這樣烹調這道菜，因為費工夫、費時間，而且賣相不夠好；但吃進嘴裏就知道其中的奧妙，一切都是值得的！

【材料】

濃蝦湯	250 毫升
伊麵	2 個
薑	3-4 片

Ingredients

- 250 ml lobster broth
- 2 E-fu noodles
- 3 to 4 slices ginger

【濃蝦湯材料】

龍蝦殼	1 隻份量
鮮蝦 （取蝦殼）	900 克
沸水 （以蓋過蝦殼為準）	1.5 公升

Lobster broth

- 1 lobster shells and head
- 900 g fresh shrimps
(pick the shrimp heads and shells)
- 1.5 litres boiling water
(enough to cover the shrimp shells and heads)

【濃蝦湯調味料】

糖	1 茶匙
鹽	半茶匙
蠔油	2 茶匙

Seasoning

- 1 tsp sugar
- 1/2 tsp salt
- 2 tsp oyster sauce

【做法】

1..... 鮮蝦去殼，留殼備用。
2..... 龍蝦殼及蝦殼放入白鑊烘香，加入沸水煲約 30 分鐘成濃蝦湯，隔渣備用。
3..... 濃蝦湯加入糖、鹽、蠔油拌勻；伊麵汆水。
4..... 燒熱鑊，倒入部分濃蝦湯，加入油、伊麵、薑片，炒至伊麵乾身，再分 2 至 3 次加入餘下之濃蝦湯，炒至伊麵入味及乾身即成（每次需炒至伊麵乾身才加入濃蝦湯；湯的份量因應口味濃淡及伊麵的韌度作適度的微調）。

Method

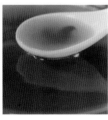

1. Shell all shrimps. Set aside the shrimp heads and shells. (Use the shelled shrimps in other recipes.)

2. Toast the lobster and shrimp shells and head in a dry wok until lightly browned and dry. Add boiling water. Cook for 30 minutes. Strain and set aside the lobster broth.

3. Season the lobster broth with sugar, salt and oyster sauce. Mix well and set aside. Blanch the E-fu noodles in boiling water until soft. Drain and set aside.

4. Heat a wok and pour in some lobster broth. Add oil, E-fu noodles and ginger. Stir-fry until all broth is absorbed by the noodles. Then add half or one-third of the remaining broth at a time. Cook until the noodles pick up all broth after each addition. Serve. (I stir-fry the noodles till dry before adding more lobster broth. That's how I fine-tune the seasoning and the consistency of the noodles to my preference.)

阿爺廚房──消失的味道

阿爺秘技 ——————— Cooking tips

- 龍蝦殼煲成濃蝦湯；龍蝦肉可用於「爆炒龍蝦球」（做法參考 p.100），一料兩吃。
- 熬蝦湯前，先用白鑊烘香蝦殼，烘得越乾蝦味越濃。
- 濃蝦湯分數次加入伊麵內，讓伊麵慢慢吸收蝦湯之精華。
- 炒伊麵時若感到黏鑊，可在鑊邊加入適量油。

- You can make lobster broth with the lobster head and shell. The lobster meat can be used to make lobster tail quick-fried with assorted vegetables (recipe on p.100). One ingredient is used to make two dishes.

- To make the lobster broth, make sure you toast the lobster and shrimp heads and shells until dry and browned. The drier they are, the more intense the flavour would be.

- Do not add the lobster broth to the E-fu noodles in one go. Add the broth in several batches to allow time for the noodles to pick up the flavours.

- If the noodles turn sticky during the stir-frying process, drizzle some oil along the rim of the wok.

爆炒龍蝦球

~Lobster tail quick-fried with assorted vegetables~

菜式特點

「爆炒」是不停地快速炒勻材料的烹調方法。今次選用了龍蝦肉、鮮百合及白玉木耳等，快速拌炒之下，能吃出各款材料的爽脆口感。

【材料】

龍蝦肉_____1 隻
　（約 1.2 公斤）
鮮百合_____90 克
白玉木耳_____70 克
　（已浸軟）
荷蘭豆_____90 克
甘筍_____20 克
薑_____3-4 片

【龍蝦肉調味料】

糖_____1 茶匙
鹽_____1 茶匙
胡椒粉_____少許
油_____1 茶匙

【調味料】

糖_____1 茶匙
鹽_____1 茶匙
蠔油_____1 湯匙

Ingredients

- 1 shelled lobster tail (about 1.2 kg)
- 90 g fresh lily bulbs
- 70 g white jade wood ear fungus (soaked in water till soft)
- 90 g snow peas
- 20 g carrot
- 3 to 4 slices ginger

Marinade for lobster tail

- 1 tsp sugar
- 1 tsp salt
- ground white pepper
- 1 tsp oil

Seasoning

- 1 tsp sugar
- 1 tsp salt
- 1 tbsp oyster sauce

【做法】

1..... 白玉木耳切塊；甘筍切小塊；荷蘭豆去筋。
2..... 龍蝦肉切斜件，加入調味料略拌。
3..... 於油鑊加入龍蝦肉拉油（約三成熟），盛起隔油備用。
4..... 起油鑊，下荷蘭豆拉油，盛起隔油備用。
5..... 起油鑊，爆香薑片，下甘筍、鮮百合翻炒，加入白玉木耳、龍蝦肉炒勻，下糖、鹽、蠔油調味，最後加入荷蘭豆炒勻即成。

Method

1. Cut the white jade wood ear into chunks. Cut carrot into small pieces. Tear off the tough veins on the snow peas.

2. Cut the shelled lobster tail diagonally into pieces. Add marinade and mix well.

3. Blanch the lobster meat in warm oil until medium-rare done. Drain and set aside.

4. Heat a wok and add oil. Blanch the snow peas. Drain and set aside.

5. Heat a wok and add oil. Stir-fry ginger, carrot and lily bulbs until fragrant. Add white jade wood ear fungus and lobster meat. Toss well. Season with salt, sugar and oyster sauce. Add snow peas and toss again. Serve.

阿爺秘技 ——————————— Cooking tips

- 龍蝦起肉後浸冰水，令龍蝦肉快速收縮，肉質爽口、彈牙。
- 龍蝦肉拉油至三成熟，可保持蝦肉外形及鎖緊肉汁，令肉質嫩滑。
- I soak the lobster tail in ice water after shelling it. This would make the muscle fibre contract, so that it would turn out springy in texture.
- Blanch the lobster meat in warm oil until medium-rare done the first time around. That would seal in the juices and secure its shape, while keeping the meat moist and tender.

海膽炒飯

~Sea urchin fried rice~

菜 式 特 點

在很早時期，香港已有海膽出產，由於漁民近海得食，很多時候都可以品嚐海膽。今次以海膽炒飯，加入蛋白和少許鹽帶出海膽的鮮味。

【材料】

新鮮海膽（一板）＿＿100 克
蛋白＿＿＿＿＿＿＿＿＿80 克
白飯＿＿＿＿＿＿＿＿＿200 克
葱花＿＿＿＿＿＿＿＿＿適量
鹽＿＿＿＿＿＿＿＿＿＿適量

Ingredients

- 100 g fresh sea urchin roe (1 pack)
- 80 g egg whites
- 200 g steamed rice
- finely chopped spring onion
- salt

【做法】

1..... 蛋白拂勻。
2..... 起油鑊，下蛋白快炒，調至小火，加入白飯，
　　　再調回大火，炒至飯粒散開，下鹽調味，加入
　　　海膽推勻及收乾，上碟，最後灑入葱花即成。

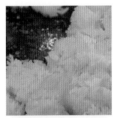
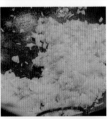

Method

1. Whisk the egg whites.
2. Heat a wok and add oil. Stir-fry egg whites quickly. Turn to low heat. Put in the rice. Turn the heat up to high and toss the rice until rice grains are separated from each other. Season with salt. Put in the sea urchin roe. Toss to mix well and till dry. Save on a serving plate. Sprinkle with finely chopped spring onion. Serve.

海膽 Sea urchin

阿爺秘技 ——————————— Cooking tips

- 先把蛋白炒勻後才加入白飯,令蛋白炒成粒粒狀,增加賣相及口感。
- 海膽分兩次下鑊炒勻,既可以避免飯粒過濕,也可以保持海膽的鮮味。
- 下鹽後才加入海膽,否則海膽會溶掉。

■ I stir-fry the egg white first before putting in the steamed rice. The egg white will be scrambled into small bits that way, looking great and adding interesting mouthfeel.

■ Add half of the sea urchin roe to the rice each time. Then toss till dry before adding the rest. That would prevent the rice from getting too wet while retaining the umami of sea urchin roe.

■ Put in the sea urchin after seasoning the rice with salt. Otherwise, the sea urchin will break down and turn gooey.

海膽蒸蛋白

~Steamed egg white custard with sea urchin roe~

菜式特點

將海膽配上蛋白一起蒸，利用最簡單的方法令海膽的鮮味盡現。蒸蛋白的關鍵是蛋白與涼開水的比例，吃到的是啖啖鮮甜的海膽，以及極至嫩滑的蛋白。

【材料】

新鮮海膽（半板）＿50 克
蛋白＿＿＿＿＿＿＿150 克
涼開水＿＿＿＿＿120 毫升

Ingredients

- 50 g fresh sea urchin roe (1/2 pack)
- 150 g egg whites
- 120 ml cold drinking water

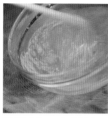

【做法】

1.....蛋白拂勻，過篩。
2.....蛋白加涼開水拌勻，盛於蒸碟內。
3.....排入海膽，用耐熱保鮮紙包好，蒸約 20 分鐘即成。

Method

1. Whisk the egg whites. Pass them through a wire mesh.
2. Add cold drinking water to the egg whites. Mix well. Save in a steaming dish.
3. Arrange the sea urchin evenly in the egg white mixture. Cover in microwave-safe cling film. Steam for 20 minutes. Serve.

阿爺廚房——消失的味道

阿爺秘技 ———————— Cooking tips

- 蛋白與涼開水的比例為 1：0.8，因海膽含有水分。
- 蛋白要過密篩才細緻、幼滑。
- 蛋白拂打後不要灑入鹽調味，否則蛋白會不滑，而且海膽會釋出水分。
- 蒸蛋白時蓋上耐熱保鮮紙，以防蒸氣倒流。

- The ratio between the egg whites and cold drinking water should be 1 to 0.8 by volume. That's a bit less water than usual as water would ooze out of the sea urchin roe and dilute the egg whites.
- Pass the whisked egg whites through a wire mesh to remove the stringy bits in the egg whites. The custard would turn out velvety that way.
- After whisking the egg whites, do not season with salt. Otherwise, the custard would be hard instead of jiggly. The salt would also draw the water out of the sea urchin roe.
- Cover the dish with cling film before steaming. That would stop the condensation from dripping back on the custard and forming blisters.

鮑魚綠豆沙

~Abalone mung bean sweet soup~

菜式特點

塔門五、六十年代的收入來源大部分來自鮑魚，當中外觀略次的鮑魚未能出售，而漁民也不想浪費，所以用來製作「鮑魚綠豆沙」，塔門人也喜歡用西米棟取代海帶，這款糖水實在是一種消失的味道。

【材料】

鮑魚仔＿＿＿＿＿600 克
　（14 隻左右）
綠豆＿＿＿＿＿＿100 克
眉豆＿＿＿＿＿＿30 克
西米棟＿＿＿＿＿40 克
陳皮＿＿＿＿＿＿1 角
生薏米＿＿＿＿＿20 克
熟薏米＿＿＿＿＿10 克
片糖＿＿＿＿＿＿1 塊
水＿＿＿＿＿＿＿3 公升

Ingredients

- 600 g baby abalones
  (about 14 abalones)
- 100 g mung beans
- 30 g black-eyed beans
- 40 g Ximidong (dried seaweed)
- 1 piece aged tangerine peel
- 20 g raw Job's tears
- 10 g puffed Job's tears
- 1 raw cane sugar slab
- 3 litres water

【做法】

1..... 生薏米浸過夜；綠豆預早 2 小時浸泡。
2..... 西米棟預早 30 分鐘浸發至軟身，洗淨。
3..... 鮑魚擦淨，汆水，起肉、去除內臟，備用。
4..... 水煮沸，下綠豆、眉豆、西米棟、陳皮、生熟薏米，煲約 1.5 小時至綠豆開花。
5..... 調至中火，下片糖調味，最後加入鮑魚用細火煲約半小時即可。

Method

1. Soak the raw Job's tears in water overnight. Drain before using. Soak mung beans in water for 2 hours.
2. Soak Ximidong in water for 30 minutes until soft. Rinse and drain.
3. Scrub the abalones clean. Blanch in boiling water. Shell them and remove the innards.
4. Boil water in a pot. Put in mung beans, black-eyed beans, Ximidong, aged tangerine peel, raw and puffed Job's tears. Boil for 1.5 hours until the mung beans start to break down and turn mushy.
5. Turn to medium heat. Add raw cane sugar slab. Put in the abalones at last. Cook over low heat for 30 minutes. Serve.

阿爺秘技 ———————— Cooking tips

- 用西米棟取代海帶、臭草，有清熱解毒之功效。
- 選用細隻的鮮鮑魚為佳。
- 加入鮑魚後，調至細火煲，可保持鮑魚軟滑的口感。
- 眉豆令綠豆沙更香、更綿。

左邊是已浸之西米棟；
右邊是乾西米棟。
Soaked Ximidong (left);
dried Ximidong (right).

- Fishermen use Ximidong in place of kelp and dried common rue. Chinese herbalists believe Ximidong clears Heat and detoxifies.

- For this recipe, small live abalones are preferred.

- After putting the abalones in, turn down the heat to low. That would help retain the tenderness of the abalones.

- Black-eyed beans make the mung bean sweet soup creamier and more fragrant.

瀨尿蝦乾蒸肉餅

~Steamed pork patty with dried mantis shrimps~

菜式特點

水上人將海產生曬保存，吃的盡是其鮮味；肉餅拌入瀨尿蝦乾及香煎鹹魚，品嚐到又香又有口感的肉餅。

【材料】

瀨尿蝦乾_____30 克
梅香鹹魚_____25 克
腩頭豬肉_____400 克
薑絲_____少許

【醃料】

胡椒粉_____少許
油_____1 茶匙
水_____3 湯匙
糖_____1 茶匙
生粉_____1 茶匙

塔門的生曬瀨尿蝦乾。
The sun-dried dried mantis shrimps in Grass Island.

Ingredients

- 30 g dried mantis shrimps
- 25 g salted fish (preferably with mushy flesh)
- 400 g pork shoulder butt
- shredded ginger

Marinade

- ground white pepper
- 1 tsp oil
- 3 tbsp water
- 1 tsp sugar
- 1 tsp potato starch

【做法】

1..... 瀨尿蝦乾用水浸軟，留起 6-8 隻，其餘切碎。
2..... 起油鑊，下梅香鹹魚煎香，拆肉後切碎。
3..... 脢頭豬肉一半切粒，另一半剁蓉。
4..... 豬肉下胡椒粉、油、部分水拌勻，再加入已切碎的瀨尿蝦乾、鹹魚肉、糖、水及生粉拌勻。
5..... 肉餅放於蒸碟內，用筷子戳開洞，於肉餅面鋪上原隻瀨尿蝦乾及薑絲，用耐熱保鮮紙包好，蒸約 18 分鐘即成。

Method

1. Soak dried mantis shrimps in water till soft. Set aside 6 to 8 of them for later use. Finely chop the rest.
2. Heat a wok and add oil. Fry the salted fish until golden. De-bone and mash the flesh.
3. Dice half of the pork. Then finely chop the rest.
4. Add ground white pepper, oil and part of the water to the pork. Mix well. Add chopped dried mantis shrimps, salted fish flesh, sugar, water and potato starch. Mix well.
5. Put the resulting pork mixture on a steaming plate and shape it like a round patty. Make a few indentations on the surface with a chopstick. Arrange the whole dried mantis shrimps and shredded ginger on top. Wrap in microwave-safe cling film. Steam for 18 minutes. Serve.

阿爺秘技 ——————— Cooking tips

- 有膏的瀨尿蝦乾味濃香；無膏的則鮮甜。
- 烹調前先浸透瀨尿蝦乾至軟身，否則肉質太硬；而且蒸熟後可減少腥味。
- 蒸肉餅必須加入適量水，讓豬肉吸收足夠水分。
- 肉餅鋪入碟後，要用筷子戳開洞以疏氣，令肉餅不會太實。

- Those dried mantis shrimps with roe in them taste rich and flavourful. Those without roe taste sweet with strong umami.
- Soak the dried mantis shrimps in water till soft before using. Otherwise, they will be too hard. Besides, they may smell fishy before cooked, but they won't taste fishy after steamed with ginger.
- When making steamed pork patty, always add some water. That would let the pork pick up the moisture and stay juicy.
- After putting the pork mixture on a steaming plate, make some indentations on the surface with a chopstick. That would allow the free circulation of steam so that the pork patty would not be too firm and stiff.

廣東蔗蝦

~Cantonese sugarcane shrimp~

菜式特點

有別於越南蔗蝦，將蔗去皮後切成扁平狀，再用加入肥
豬肉的蝦膠包住，製作出廣東版蔗蝦，口感更香、更
滑，而且吃到竹蔗清香之味。

【材料】

竹蔗	4-6 條
蝦仁	250 克
肥豬肉	60 克
冰凍啤酒	170 毫升
天婦羅炸粉	170 克

Ingredients

- 4 to 6 segments sugarcane
- 250 g shelled shrimps
- 60 g fatty pork
- 170 ml ice-cold beer
- 170 g tempura batter mix

【醃料】

糖	1 茶匙
鹽	1 茶匙
生粉	半茶匙
麻油	少許

Marinade

- 1 tsp sugar
- 1 tsp salt
- 1/2 tsp potato starch
- sesame oil

阿爺秘技 ——————— Cooking tips

- 蝦膠加入了肥豬肉，增加香味及軟滑的口感，肥豬肉與蝦膠的比例大約是 1：4。
- 天婦羅粉用冰凍啤酒調開，拌勻後再冷藏，能保持冰凍的溫度。
- 用啤酒調開的炸漿，炸物除了有啤酒之香氣，也吃到鬆化之口感。

- Adding fatty pork to minced shrimp give the dish extra flavour and tender texture. The ratio between fatty pork and shrimp is about 1 to 4.
- When you make the deep-frying batter, make sure you add ice-cold beer to the tempura mix. Mix well and then refrigerate again to keep it cold.
- The beer releases bubbles when heated so that it makes the crust airy and fluffy. It also imparts a malt-like depth and aroma.

【做法】

1..... 竹蔗去皮、切片，於中間部分用小刀切出凹位。肥豬肉切粒。
2..... 蝦仁用刀背拍扁，剁碎，打成蝦膠，加入肥豬肉拌勻，下糖、鹽、生粉、麻油調味，撻至起膠。
3..... 竹蔗拍上少許生粉，於中間凹位包上蝦膠。
4..... 冰凍啤酒與天婦羅炸粉拌勻成炸漿，再冷藏30分鐘。
5..... 蔗蝦沾上冰凍炸漿，放入油鑊內用中大火炸至金黃色即成。

Method

1. Peel the sugarcane and cut into long thin slices. Cut a notch at the centre with a paring knife. Set aside. Dice the fatty pork.

2. Crush the shelled shrimps with the flat side of a knife. Then finely chop them. Add fatty pork and mix well in a bowl. Add sugar, salt, potato starch and sesame oil. Lift the mixture off the bowl and slab it back in forcefully a few times until sticky.

3. Coat the sugarcane slices lightly in potato starch. Wrap the indentation at the centre with the shrimp and pork mixture from step 2.

4. Mix cold beer with the tempura batter mix to make a deep-frying batter. Refrigerate for 30 minutes.

5. Coat the sugarcane shrimp in the cold batter. Deep-fry in oil over medium-high heat until golden. Drain and serve.

酸菜桂花魚

~Mandarin fish fillet in fish stock with
pickled mustard greens~

菜式特點

「酸菜魚」起源於重慶，但來源眾說紛紜，至今也無從考證。
「酸菜魚」一般用鯇魚烹調，這次則選用桂花魚，以檸檬代替
醋，還加入紫仔菜、大豆芽取鹹味和甜味，達到提鮮的作用。

【材料】

桂花魚＿＿＿＿＿＿1 條
　（900 克 -1.2 公斤）
鹹酸菜＿＿＿＿＿＿100 克
大豆芽＿＿＿＿＿＿100 克
紮仔菜＿＿＿＿＿＿50 克
木耳＿＿＿＿＿＿＿30 克（已浸軟）
番茄＿＿＿＿＿＿＿1 個
檸檬片＿＿＿＿＿＿半個
檸檬汁＿＿＿＿＿＿半個份量
白胡椒粒＿＿＿＿＿5 克
乾辣椒＿＿＿＿＿＿3-4 隻
花椒油＿＿＿＿＿＿2-3 湯匙
鮮花椒＿＿＿＿＿＿20 克
薑＿＿＿＿＿＿＿＿3-5 片
唐芹＿＿＿＿＿＿＿少許

【魚片調味料】

胡椒粉＿＿＿＿＿＿少許
紹興酒＿＿＿＿＿＿1 茶匙
油＿＿＿＿＿＿＿＿1 茶匙

Ingredients

- 1 mandarin fish
  (about 900 g to 1.2 kg)
- 100 g salty pickled mustard greens
- 100 g soybean sprouts
- 50 g dried kohlrabi greens
- 30 g wood ear fungus
  (soaked in water till soft)
- 1 tomato
- 1/2 lemon (sliced)
- 1/2 lemon (juice squeezed)
- 5 g white peppercorns
- 3 to 4 dried chillies
- 2 to 3 tbsp Sichuan peppercorn oil
- 20 g fresh Sichuan peppercorns
- 3 to 5 slices ginger
- Chinese celery

Seasoning for fish

- ground white pepper
- 1 tsp Shaoxing wine
- 1 tsp oil

【做法】

1..... 桂花魚起肉，骨斬件，魚肉切片。
2..... 起油鑊，下魚骨煎香，加入沸水煲 20-30 分鐘成魚湯，隔渣備用。
3..... 魚片下胡椒粉、紹興酒、油拌勻。
4..... 鹹酸菜、紮仔菜切片；木耳切塊；唐芹切段；番茄去皮、切件。
5..... 鹹酸菜、紮仔菜汆水，再下白鑊烘至乾身，盛起備用。
6..... 起油鑊，爆香薑片，下鹹酸菜、紮仔菜、大豆芽，炒至大豆芽軟身，加入番茄炒勻，倒入魚湯、檸檬片，檸檬汁、木耳、白胡椒粒、乾辣椒煮滾，材料盛起上碟。
7..... 魚片放入魚湯內，灼熟後盛起。
8..... 花椒油下鑊煮熱，加入鮮花椒，煮至出味後淋於酸菜魚上面，再放上唐芹即成。

阿爺秘技 ——————— Cooking tips

- 用番茄及檸檬取代醋的酸味，吃出天然的味道。
- 魚骨煎湯取其鮮味；大豆芽取其甜味。
- 鹹酸菜及紮仔菜經過汆水及白鑊烘後，會減少鹹味及去掉苦澀味，散發醃菜的香氣。

- I use tomato and lemon to replace vinegar in the traditional recipe. Their natural tartness works well with the pickled and dried greens.
- Fry the fish bones and head add umami to the fish stock, while soybean sprouts impart sweetness.
- Pickled mustard greens and dried kohlrabi greens would taste less salty and bitter after being blanched in water and fried in dry wok. Their aromas would also be heightened that way.

Method

1. De-bone the fish. Chop the bones into chunks. Set aside the bones and fish head for later use. Slice the fish fillet.
2. Heat wok and add oil. Fry the fish bones and head until golden on both sides. Pour in boiling hot water. Cook for 20 to 30 minutes to make the fish stock. Strain the stock.
3. Add marinade to the sliced fish fillet. Mix well.
4. Slice the pickled mustard greens and dried kohlrabi greens. Set aside. Cut wood ear into chunks. Cut Chinese celery into short lengths. Peel and cut tomato into chunks.
5. Blanch pickled mustard greens and dried kohlrabi greens in boiling water. Drain and fry them in a dry wok until dry. Set aside.
6. Heat wok and add oil. Stir-fry ginger until fragrant. Add pickled mustard greens, dried kohlrabi greens, and soybean sprouts. Toss until the soybean sprouts start to wilt. Add tomato and toss again. Then put in fish stock, sliced lemon, lemon juice, wood ear fungus, white peppercorns and dried chillies. Bring to the boil. Save the ingredients on a serving plate.
7. Cook the sliced fish in the fish stock from step 6 until done. Save on the serving plate over the pickled mustard greens mixture.
8. Heat Sichuan peppercorn oil in a wok. Fry the fresh Sichuan peppercorns until flavours are infused. Drizzle the hot oil over the fish and mustard greens. Garnish with Chinese celery. Serve.

⊙ 示範短片

黃鱔焗飯

~Clay pot rice with rice-paddy eel~

菜 式 特 點

用原條黃鱔焗飯，是最古老的做法。從前的台山人，在煮飯途中，將黃鱔打暈及去掉表面的黏液。當飯熟了，馬上宰黃鱔，將牠的腸子扯出來後，把黃鱔圍在飯面，保留黃鱔的血，他們認為鱔血才是最補益。

【材料】

黃鱔＿＿＿＿＿＿1 條
米＿＿＿＿＿＿300 克
水＿＿＿＿＿＿300 毫升
陳皮＿＿＿＿＿＿1 個（切絲）
薑絲＿＿＿＿＿＿少許
葱花＿＿＿＿＿＿少許

【洗黃鱔料】

鹽＿＿＿＿＿＿適量
生粉＿＿＿＿＿＿適量
紹興酒＿＿＿＿＿＿適量
薑汁＿＿＿＿＿＿25 克
胡椒粉＿＿＿＿＿＿少許

【醃料】

乾葱＿＿＿＿＿＿10 克
薑蓉＿＿＿＿＿＿10 克
鹽＿＿＿＿＿＿1 茶匙
糖＿＿＿＿＿＿1 茶匙
生粉＿＿＿＿＿＿半茶匙
紹興酒＿＿＿＿＿＿少許

【煲仔飯豉油】

生抽＿＿＿＿＿＿150 毫升
糖＿＿＿＿＿＿2 茶匙
老抽＿＿＿＿＿＿1 茶匙

Ingredients

- 1 rice-paddy eel
- 300 g rice
- 300 ml water
- 1 whole piece aged tangerine peel (finely shredded)
- shredded ginger
- finely chopped spring onion

Ingredients for rubbing the eel

- salt
- potato starch
- Shaoxing wine
- 25 g ginger juice
- ground white pepper

Marinade

- 10 g shallot
- 10 g grated ginger
- 1 tsp salt
- 1 tsp sugar
- 1/2 tsp potato starch
- Shaoxing wine

Sweet soy sauce for clay pot rice

- 150 ml light soy sauce
- 2 tsp sugar
- 1 tsp dark soy sauce

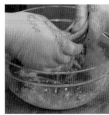

【做法】

1..... 將鹽、生粉、胡椒粉、薑汁、紹興酒拌勻，倒入黃鱔中內外搓勻，用水把黃鱔清洗乾淨。

2..... 從頭到尾在黃鱔背上每隔 1 吋切一刀，只需切斷背骨，保留魚身不斷開。

3..... 下乾葱、薑蓉、鹽、糖、生粉、紹興酒與黃鱔拌勻醃味。

4..... 砂鍋內加入米及水煲飯，中大火煮至收水後，加入原條黃鱔、陳皮絲，加蓋轉中細火焗約 12 分鐘，關火焗一會。

5..... 煮沸生抽，關火後加入糖及老抽拌勻。

6..... 飯面灑上薑絲、葱花，淋上熱油及煲仔飯豉油即成。

阿爺秘技 —— Cooking tips

- 用薑汁酒拌勻生粉洗淨黃鱔，加上用百潔布擦淨外皮，能去掉黃鱔的潺液。
- 黃鱔不宜醃味太久，否則鱔肉會散開來。
- 煲仔飯熄火後，焗煮片刻有助飯粒收水透身。

- I mix ginger juice, wine and potato starch together and rub the eel in the mixture. Then I use a scouring pad to scrub the skin. That would ensure all the slime on the eel is removed.

- Do not marinate the eel for too long. Otherwise, its flesh may break down.

- After the rice is cooked, I turn off the heat and let it sit for a while. That would allow time for the rice grains to pick up the steam and cook through.

阿爺廚房——消失的味道

Method

1. In a bowl, mix together salt, potato starch, ground white pepper, ginger juice and Shaoxing wine. Put in the eel and scrub evenly on both the inside and outside. Then rinse well with fresh water.
2. Make a light cut along the back of the eel at 1-inch intervals. Just cut through the spine without cutting all the way through.
3. Put the eel into a bowl again. Add shallot, grated ginger, salt, sugar, potato starch and Shaoxing wine. Mix well and rub the marinade on both inside and outside.
4. Put rice and water into a clay pot. Bring to the boil and cook over medium-high heat until no liquid is visible over the rice. Then coil the eel over the rice. Sprinkle with shredded aged tangerine peel. Turn to medium-low heat and cover the lid. Cook for 12 more minutes. Turn off the heat and leave it for a while.
5. To make the sweet soy sauce, bring light soy sauce to the boil. Turn off the heat. Stir in sugar and dark soy sauce. Mix well.
6. Sprinkle with shredded ginger and chopped spring onion over the rice. Drizzle with smoking hot oil and the sweet soy sauce from step 5. Serve.

銀芽肉絲炒鱔糊

~Stir-fried eel with bean sprouts and shredded pork~

菜式特點

「銀芽炒鱔糊」是一道上海人家都會做的家常菜，今次加入肉絲製作成「銀芽肉絲炒鱔糊」，吃起來多了一點肉香味，絕對惹味！

【材料】

黃鱔＿＿＿＿＿＿3 條
銀芽＿＿＿＿＿＿200 克
豬柳＿＿＿＿＿＿160 克
韭黃＿＿＿＿＿＿50 克
青甜椒＿＿＿＿＿50 克
紅甜椒＿＿＿＿＿50 克
黃甜椒＿＿＿＿＿50 克
蒜蓉＿＿＿＿＿＿15 克
豬背肥肉＿＿＿＿80 克

Ingredients

- 3 rice-paddy eels
- 200 g mung bean sprouts
- 160 g pork loin
- 50 g yellow chives
- 50 g green bell pepper
- 50 g red bell pepper
- 50 g yellow bell pepper
- 15 g grated garlic
- 80 g fatty pork (back cut)

【豬柳調味料】

乾葱＿＿＿＿＿＿2 粒
蒜蓉＿＿＿＿＿＿1 湯匙
糖＿＿＿＿＿＿＿半茶匙
鹽＿＿＿＿＿＿＿少許
紹興酒＿＿＿＿＿少許
油＿＿＿＿＿＿＿少許

Seasoning for pork loin

- 2 shallots
- 1 tbsp grated garlic
- 1/2 tsp sugar
- salt
- Shaoxing wine
- oil

【黃鱔調味料】

薑汁＿＿＿＿＿＿1 湯匙
紹興酒＿＿＿＿＿1 茶匙
糖＿＿＿＿＿＿＿1 茶匙
鹽＿＿＿＿＿＿＿半茶匙

Seasoning for eels

- 1 tbsp ginger juice
- 1 tsp Shaoxing wine
- 1 tsp sugar
- 1/2 tsp salt

【做法】

▶ 示範短片

1..... 黃鱔起肉、斜切成絲;乾葱切蓉;甜椒切絲。

2..... 豬柳切絲,下乾葱、蒜蓉、糖、鹽、紹興酒、油拌勻。

3..... 黃鱔下薑汁、紹興酒、糖、鹽拌勻。

4..... 肥豬肉切粒;銀芽下白鑊烘至乾身。

5..... 燒熱白鑊,下肥豬肉以慢火炸出豬油,隔渣備用。

6..... 黃鱔沾上薄薄一層生粉。起油鑊,下黃鱔拉一拉油,盛起隔油。

7..... 起油鑊,下豬柳絲炒至半熟,加入甜椒炒勻,放入黃鱔,下適量糖、鹽、老抽調味,加入銀芽、韭黃炒勻,上碟。

8..... 於碟內炒鱔糊中心留出空位,放上蒜蓉。燒熱豬油,趁熱潷入蒜蓉內即成。

阿爺秘技 ————— Cooking tips

- 銀芽先放入白鑊烘乾,可減少炒時釋出水分。

- 黃鱔上粉後要盡量攤開,否則糊成一團;盡量將生粉包着鱔肉,否則脆度不足。

- 黃鱔拉油時,要輕輕撥動,以免鱔肉黏在一起。

- 熱豬油潷入蒜蓉,能夠提升整道菜之蒜香。

- Toast mung bean sprouts in a dry wok first. Otherwise, the stir-fried eel will turn out watery.

- After you coat the eel strips in potato starch, try your best to separate them from one and other. Otherwise, they would clump together and turn into one big blob after fried. Try to coat every strip evenly in potato starch, so that they turn out crispy enough.

- When you blanch the eel strips in oil, gently wiggle them to separate them, so that they won't clump together.

- I drizzle hot lard over raw grated garlic at last. That would boost the garlicky aroma in major ways.

Method

1. De-bone the eels. Cut into fine strips diagonally. Finely chop the shallots. Shred the bell peppers.
2. Shred the pork loin. Add shallots, garlic, sugar, salt, Shaoxing wine and oil in this particular order. Mix well.
3. Put eel strips into a bowl. Add ginger juice, Shaoxing wine, sugar and salt. Mix well.
4. Dice the fatty pork. Toast the mung bean sprouts in a dry wok until soft.
5. Heat a dry wok and put in the fatty pork. Fry over low heat to render the lard. Strain the lard.
6. Coat the eel strips in potato starch lightly. Heat wok and add oil. Put the eel strips into a strainer ladle. Deep-fry briefly. Drain.
7. Heat wok and add oil. Put in the pork loin and stir-fry until half-cooked. Put in the bell peppers. Toss well. Then add the eel strips. Season with sugar, salt and dark soy sauce. Add mung bean sprouts and yellow chives. Toss well. Save on a serving plate.
8. Make a well at the centre of the eel mixture. Put in grated garlic. Heat the lard from step 5. Drizzle into the well to cook the garlic. Serve.

鹹檸檬浸黃腳鱲

~Yellowfin sea bream poached in salted lemon broth~

菜式特點

用鹹檸檬、話梅及陳皮浸煮黃腳鱲，加上在魚腹內輕輕剔開魚肉，令魚肉受熱均勻，吃到的是滑溜的魚肉。

【材料】

黃腳鱲_____2 條
　（每條約重 450 克）
鹹檸檬_____2 個
話梅_____2 粒
陳皮_____2 角
青檸檬_____1/2 個
黃檸檬_____1/4 個（榨汁）
指天椒圈_____適量

Ingredients

- 2 yellowfin sea breams (about 450 g each)
- 2 salted lemon
- 2 dried liquorice plums
- 2 pieces aged tangerine peel
- 1/2 lime
- 1/4 lemon (juice squeezed)
- bird's eye chillies (cut into rings)

【做法】

1..... 於黃腳鱲的腹肉骨位輕輕切開，洗淨、抹乾。

2..... 鹹檸檬輕輕切開，有助散發香味；青檸檬取皮、切絲。

3..... 陳皮浸軟，備用。

4..... 話梅用凍滾水浸軟，取 1 粒搗碎。

5..... 煮滾水，下鹹檸檬、原粒話梅、陳皮，煮沸。下少許油及黃腳鱲，滾後加蓋，關火焗 5 分鐘。

6..... 黃腳鱲浸熟後上碟，淋上檸檬汁，少許鹽調味，淋上滾油，放上話梅蓉、青檸皮、紅椒圈，最後淋上少許湯即成。

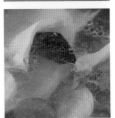

Method

1. Cut the flesh along the spine of the fish on the inside. Rinse and wipe dry.
2. Make a couple cuts on the salted lemon to help the fragrance infuse. Set aside. Peel the lime and use the rind only. Finely shred the lime rind.
3. Soak the aged tangerine peel in water till soft. Set aside.
4. Soak the dried liquorice plums in cold drinking water till soft. De-seed one of them and crush to mash the flesh. Leave the other one in whole.
5. Boil water in a wok. Put in salted lemons, the whole dried liquorice plum and the aged tangerine peel. Bring to the boil again. Add a dash of oil and put in the fish. Bring to the boil. Turn off the heat and cover the lid. Leave the fish in the hot broth for 5 minutes.
6. Transfer the fish onto a serving plate. Drizzle with lemon juice. Sprinkle with a pinch of salt. Drizzle with smoking hot oil and arrange mashed plum, lime rind and bird's eye chillies over. Ladle some broth on top. Serve.

 _____ Cooking tips

- 水大滾後關火，用餘溫浸熟黃腳鱲，令魚肉保持嫩滑。
- 用話梅蓉襯托鹹檸檬，更能突出鮮味。
- After putting the fish in the broth, bring to a vigorous boil and turn off the heat. Let the residual heat cook through the fish gently.
- The dried liquorice plums complement the tartness of salted lemon. They work together to accentuate the umami of the fish.

紅炆龍躉腩

~Red braised giant grouper belly~

菜式特點

「紅炆龍躉腩」先在魚的厚肉處切開，讓它平均地煮熟，基本調味之後抹上生粉泡油，之後把冰糖煮至焦糖色再炆煮，上桌時令菜餚呈偏紅色澤，此為紅炆的烹調方法。

【材料】

龍躉腩＿＿＿＿＿1.35 公斤
　（約 2.25 斤）
獨子蒜＿＿＿＿＿180 克
薑片＿＿＿＿＿＿70 克
冬菇＿＿＿＿＿＿185 克（已浸發）
雲耳＿＿＿＿＿＿120 克（已浸發）
葱段＿＿＿＿＿＿35 克
乾葱＿＿＿＿＿＿3 粒
火腩＿＿＿＿＿＿170 克（切件）
冰糖＿＿＿＿＿＿100 克（煮成焦糖）
紹興酒＿＿＿＿＿2 湯匙

【龍躉腩醃料】

糖＿＿＿＿＿＿＿2 茶匙
胡椒粉＿＿＿＿＿1 茶匙
鹽＿＿＿＿＿＿＿2 茶匙
紹興酒＿＿＿＿＿1 湯匙
薑汁＿＿＿＿＿＿4 湯匙

【調味料】

冰糖＿＿＿＿＿＿20 克
老抽＿＿＿＿＿＿20 克
生抽＿＿＿＿＿＿40 克
鹽＿＿＿＿＿＿＿1 茶匙
蠔油＿＿＿＿＿＿50 克

Ingredients

- 1.35 kg giant grouper belly
- 180 g solo garlic
- 70 g sliced ginger
- 185 g dried shiitake mushrooms (soaked in water till soft)
- 120 g dried cloud ear fungus (soaked in water till soft)
- 35 g spring onion (cut into short lengths)
- 3 shallots
- 170 g roast pork belly (cut into chunks)
- 100 g rock sugar (for making caramel)
- 2 tbsp Shaoxing wine

Marinade for fish

- 2 tsp sugar
- 1 tsp ground white pepper
- 2 tsp salt
- 1 tbsp Shaoxing wine
- 4 tbsp ginger juice

Seasoning

- 20 g rock sugar
- 20 g dark soy sauce
- 40 g light soy sauce
- 1 tsp salt
- 50 g oyster sauce

【做法】

1..... 在龍躉腩厚肉位置輕切數刀，下糖、胡椒粉、鹽抹勻魚身，再下紹興酒、薑汁塗抹魚身備用。

2..... 龍躉腩沾上生粉，備用。

3..... 取約 1/3 薑片及獨子蒜，下油鑊炸至金黃色及熟透，盛起隔油備用。

4..... 起油鑊，龍躉腩拉油至兩面金黃色，盛起隔油備用。

5..... 起油鑊，爆香生薑片、乾葱、獨子蒜及火腩，加入約 1-1.2 公升水及紹興酒煲滾，加入熟薑、熟蒜、冬菇、冰糖、生抽、老抽、鹽拌勻，盛起一半材料，放入龍躉腩後，倒回剛才盛起的材料鋪在龍躉腩上，加水蓋至龍躉腩約 8 成位置。

6..... 用另一白鑊，下冰糖煮溶，直至融化成焦糖溶液，倒入正在炆的龍躉腩鍋內。

7..... 待翻滾後，加入蠔油調味，每邊炆約 30 分鐘，令兩面上色均勻，共約 1 小時。

8..... 加入已浸軟雲耳煮 3-4 分鐘，保持爽口；如喜歡雲耳腍滑，可炆煮長一點時間。

9..... 龍躉腩炆至入味並收汁後，加入葱段，熄火，焗約 2 分鐘即成。

▶ 示範短片

Method

1. Make a few incisions on the fleshiest part of the giant grouper belly. Add sugar, ground white pepper and salt. Rub evenly all over the fish. Add Shaoxing wine and ginger juice. Rub evenly.

2. Coat the fish in potato starch.

3. Heat wok and add oil. Put in 1/3 of the sliced ginger and solo garlic to fry until golden and cooked through. Remove and drain excessive oil.

4. In the same wok, heat oil and deep-fry the fish until both sides golden. Set aside and drain excessive oil.

5. Heat wok and add oil. Stir-fry the remaining sliced ginger and solo garlic. Add shallots and roast pork belly. Pour in about 1 to 1.2 liters of water and Shaoxing wine. Bring to the boil. Add fried ginger, fried solo garlic, shiitake mushrooms, rock sugar, light and dark soy sauce, and salt. Mix well. Set aside half of the mixture. Put in the fried fish and put the mixture back over the fish. Add water up to about 80% of the height of the fish.

6. In a dry wok, heat rock sugar until it melts and caramelized. Pour this caramel over the fish in the other wok.

7. Heat until it comes to a boil again. Add oyster sauce to season. Braise for 30 minutes. Flip to braise the other side for 30 minutes to colour both sides evenly. That means the fish has been cooked for 1 hour in total.

8. Add the rehydrated cloud ear fungus. If you prefer the fungus to be crunchy, cook for 3 to 4 minutes. If you want the fungus gooey and gelatinous, cook for longer to your desired consistency.

9. Cook the fish till it picks up the seasoning and the sauce reduces. Add spring onion and turn off the heat. Cover the lid and wait for 2 minutes. Serve.

阿爺秘技 ————————— Cooking tips

- 在龍躉腩厚肉位置輕切，烹調時更易入味。
- 龍躉腩撲上生粉後拉油，炆煮時較耐火。
- Making light incisions on the fleshiest part of the fish helps the seasoning penetrate and makes all parts of the fish cooked at roughly the same time.
- Coat the fish in potato starch and deep-fry it in oil before braising. That would help the fish hold its shape throughout the prolonged cooking process.

番茄花蛤湯

~Tomato clam soup~

菜式特點

花蛤容易藏有泥沙，先將花蛤氽水，能吐淨泥沙之餘，也可去掉沒張口之花蛤，確保進食衛生。番茄湯加入大頭菜及鹹蛋白提鮮，令湯水帶有鹹鮮之香味。

【材料】

番茄＿＿＿＿＿＿＿300 克
花蛤＿＿＿＿＿＿＿600 克
薑＿＿＿＿＿＿＿＿50 克
糖＿＿＿＿＿＿＿＿1 茶匙
鹹蛋白＿＿＿＿＿＿30 克
大頭菜＿＿＿＿＿＿5 克
葱＿＿＿＿＿＿＿＿10 克
水＿＿＿＿＿＿＿＿2 公升

【做法】

1..... 花蛤清洗乾淨，汆水至剛開口吐沙，撈起備用。
2..... 番茄切件；薑、大頭菜切絲；葱切成葱花；鹹蛋白拂勻，備用。
3..... 湯鍋內注入 2 公升清水，煲滾後加入薑絲、大頭菜絲、番茄及糖，大火煲 5 分鐘。
4..... 待番茄變軟身，加入花蛤煲 10 分鐘，見花蛤完全開口，熄火，倒入鹹蛋白拌勻，灑上葱花即可。

Ingredients

- 300 g tomatoes
- 600 g clams
  (Venus shells)
- 50 g ginger
- 1 tsp sugar
- 30 g salted egg white
- 5 g salted kohlrabi
- 10 g spring onion
- 2 litres water

Method

1. Rinse the clams. Blanch in boiling water until just begin to open. Drain and set aside.
2. Cut tomatoes into pieces. Finely shred ginger and salted kohlrabi. Finely chop spring onion. Whisk the salted egg white.
3. Boil 2 litres of water in a pot. Put in shredded ginger and salted kohlrabi, tomatoes and sugar. Boil over high heat for 5 minutes.
4. Cook till tomatoes turn soft. Add clams and cook for 10 minutes until the clams open completely. Turn off the heat. Stir in salted egg white. Sprinkle with chopped spring onion. Serve.

阿爺秘技　　　　　　　Cooking tips

- 加入大頭菜可提升花蛤湯的鮮味。
- 番茄切成小件，更容易散發番茄香味。
- Adding salted kohlrabi helps elevate the umami of the clam soup.
- Cut the tomatoes into small pieces so that the flavours are infused more readily into the soup.

魚湯鮮蝦銀針粉

~Silver needle noodles with shrimps in fish stock~

菜式特點

銀針粉又名老鼠粉，是一種客家傳統麵食，起源於廣東
梅州大埔一帶。早年的銀針粉是由家庭主婦逐條搓成，
有大有小、有粗有細，現時則以機器製造出來。

【材料】

雜魚仔	600 克
銀針粉	300 克
鮮蝦	10 隻
肥豬肉	600 克
唐芹	30 克
韭黃	3 條
雞蛋	2 個（拂勻）
薑片	30 克（略拍扁）

Ingredients

- 600 g assorted small fish
- 300 g silver needle noodles
- 10 shrimps
- 600 g fatty pork
- 30 g Chinese celery
- 3 sprigs yellow chives
- 2 eggs (whisked)
- 30 g ginger (crushed, sliced)

【做法】

1..... 鮮蝦洗淨、去殼，挑走蝦腸備用。

2..... 韭黃及唐芹洗淨，切段備用。

3..... 銀針粉過熱水，略去掉油分，瀝乾水分。

4..... 雜魚仔洗淨，抹乾表面水分。

5..... 薑片放入已加油的鑊中爆香，放入雜魚以中火煎香兩邊魚身至金黃色，以大滾水撞落鑊內，大火煲滾後再轉中火煲45分鐘，備用。

6..... 肥豬肉洗淨，切碎，放入白鑊用小火炸出豬油，待豬油渣浮起，灒少量清水加蓋，水分消失後盛起備用。

7..... 在平底鍋抹上一層薄薄的油，以小火熱鍋，加入蛋液搖晃鍋子使蛋液鋪成薄薄一層，略凝固後，將鍋子離火約20-30秒降溫。

8..... 待溫度微降後，蛋皮邊微微翹起來，這時可反面煎另一面。蛋皮放涼，捲起切幼絲備用。

9..... 魚湯隔去湯料，注入鍋內，加入銀針粉略滾，盛於碗內。

10... 魚湯滾後加入鮮蝦，煮滾後以鹽調味，盛起鮮蝦放在碗內，加入韭黃、唐芹，淋上魚湯，放上豬油渣及蛋絲即可。

Method

1. Rinse the shrimps and shell them. Devein and set aside.
2. Rinse the yellow chives and Chinese celery. Cut into short lengths.
3. Rinse the silver needle noodles with hot water to remove excessive oil. Drain well.
4. Rinse the fish and dress well. Wipe dry.
5. Heat wok and add oil. Fry the ginger until fragrant. Then put in the fish and fry until golden on both sides. Pour in vigorously boiling water. Bring to the boil over high heat. Then turn to medium heat and cook for 45 minutes.
6. Rinse the fatty pork and chop it. Fry in a dry wok over low heat to render the lard. When the cracklings float, drizzle with a dash of water and cover the lid immediately. Cook until the sizzling sound fades off and all water evaporates. Set aside the lard (to be used in other recipes) and the cracklings separately.
7. Smear a thin layer of oil in a pan. Heat the pan over low heat. Pour in the whisked eggs and swirl the pan to coat the bottom evenly. Cook until half set. Remove the pan from the heat for 20 to 30 seconds to cool it off slightly.
8. Wait till the pan is cooler and the edge of the omelette starts to curl up. Flip the omelette upside down to fry the other side. Set aside to let cool. Roll it up and shred it finely.
9. Strain the fish stock from step 5 and save in another pot. Put in the silver needle noodles and bring to a gentle simmer. Save the noodles into a serving bowl.
10. Boil the stock again. Put in the shrimps and bring to the boil again. Season with salt. Save in the serving bowl over the bed of silver needle noodles. Arrange the yellow chives and Chinese celery on top. Drizzle with the hot fish stock. Garnish with cracklings from step 6 and shredded omelette from step 8. Serve.

阿爺秘技 ——————— Cooking tips

- 待豬油渣炸至七、八成時潛水，令豬油渣更鬆化可口。
- 魚煎香後加入滾水煲煮，湯色呈現奶白，入口香滑。
- When you render the lard and make cracklings, fry the pork till it's almost cooked through. Then drizzle with water to make the cracklings extra puffy and crispy.
- Make sure you fry the fish until golden and browned before pouring vigorously boiling water in. That's the key to creamy white stock.

蔬菜鮮魷麵

~Fried noodles with squid and vegetables~

菜式特點

這次用大尾魷來炒麵，鮮味十足、入口爽脆，細心地將大尾魷切成細絲，享用時與麵條一起夾來吃，滋味無窮！

【材料】

大尾魷＿＿＿＿＿100 克
椰菜＿＿＿＿＿40 克
甘筍＿＿＿＿＿30 克
上海幼麵＿＿＿＿150 克
蝦醬＿＿＿＿＿30 克
糖＿＿＿＿＿1 湯匙
薑粒＿＿＿＿＿適量
乾葱蓉＿＿＿＿適量
紅辣椒絲＿＿＿適量

Ingredients

- 100 g oval squid
- 40 g cabbage
- 30 g carrot
- 150 g Shanghainese fine noodles
- 30 g fermented shrimp paste
- 1 tbsp sugar
- diced ginger
- chopped shallots
- shredded red chillies

阿爺秘技 — Cooking tips

- 大尾魷又稱為荷包魷，肉質爽脆、鮮味重。
- 鮮魷魚絲幼度如上海幼麵，與其他材料搭配，享用時口感佳。
- 鮮魷魚絲最後才加入炒煮，炒至八、九成上碟最適宜；炒得太久肉質變韌。
- Oval squid is flavourful and crisp in texture.
- I shred the squid into about the same thickness as the noodles and other ingredients. That would ensure a consistent texture and mouthfeel.
- I put in the squid at last and only toss it till they are slightly undercooked (medium-well done). The squid tends to be chewy and rubbery if cooked thoroughly.

【做法】

1..... 椰菜、甘筍切成長幼絲。
2..... 大尾魷去衣，洗淨，切幼條（粗度如上海幼麵）。
3..... 上海麵氽水，浸於冰水待一會，隔水備用。
4..... 起油鑊，下薑粒及乾葱蓉，炒香蝦醬拌勻，放入上海麵拌開，下椰菜絲及甘筍絲炒勻。
5..... 加入鮮魷魚絲，灑入糖及紅椒絲，炒勻上碟即成。

Method

1. Finely shred cabbage and carrot.
2. Peel and discard the purple skin of the squid. Rinse and cut into fine strips (about the same thickness as the noodles).
3. Blanch the noodles in boiling water. Soak in ice water briefly. Drain and set aside.
4. Heat wok and add oil. Put in diced ginger and shallots. Stir-fry fermented shrimp paste to mix well. Put in the blanched noodles and toss to separate each strand of noodles. Add cabbage and carrot. Toss to mix well.
5. Put in the squid. Sprinkle with sugar and red chillies. Toss well and serve.

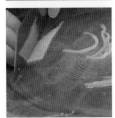

第三章

蔬菜、豆類

粟米、豆漿、甜豆等看似簡單的材料，在鼎爺細膩的刀工及技巧下，一道道精巧的菜式登場，例如百花釀甜豆、豆香餅、豆腐煎蛋等，你會讚嘆他如表演般的烹調過程，帶你進入由視覺與味覺獲得的雙重享受。

椒鹽粟米

~Deep-fried sweet corn in five-spice salt~

菜式特點

以前，人們會把粟米當主糧，五穀豐登時必會有粟米。
現在物質豐富，粟米不再是現代人的主糧了。今次先將
粟米蒸熟，再拿去香炸，搭配椒鹽品嚐，變成香口惹味
的「椒鹽粟米」。

材料

黃粟米	1 條
白粟米	1 條
蒜頭	3 粒
五香粉	1 茶匙
幼鹽	2 茶匙
葱	適量
紅尖椒	半條

調味料

糖	1 茶匙
鹽	1 茶匙

Ingredients

- 1 ear yellow corn
- 1 ear white corn
- 3 cloves garlic
- 1 tsp five-spice powder
- 2 tsp table salt
- spring onion
- 1/2 red chilli

Seasoning

- 1 tsp sugar
- 1 tsp salt

▶ 示範短片

做法

1..... 粟米洗淨，抹乾水分，原條粟米切成兩段，再將每段粟米垂直一開四，每塊削去部分粟米芯，切成長塊狀，隔水蒸 10 分鐘。

2..... 葱洗淨，切段及切粒；蒜頭切片；紅尖椒切圈。

3..... 白鑊炒香幼鹽，盛起後加入五香粉拌勻，成為椒鹽備用。

4..... 鍋內加油，放入粟米半煎炸至金黃色並呈微彎狀態，盛起備用。

5..... 鍋內剩下足夠的油，下蒜片及紅椒圈炒勻，加入糖、鹽調味，加入已拉油的粟米炒勻，灑入椒鹽拌勻，最後加入葱段及葱花炒勻即成。

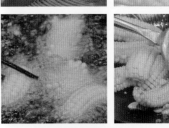
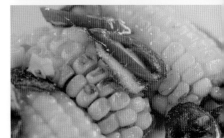

Method

1. Rinse the sweet corn. Wipe dry. Cut each ear of corn in half across the length. Then cut each lengthwise into quarters. Cut off part of the cob. Steam them for 10 minutes.

2. Rinse the spring onion. Cut some into short lengths and finely chop the rest. Slice garlic. Cut red chilli into rings.

3. Heat table salt in a dry wok until fragrant. Save in a bowl and add five-spice powder. Mix well and set aside. It is the five-spice salt.

4. Heat oil in a wok. Semi-deep fry the steamed sweet corn until they curl up. Drain.

5. Save some oil in the wok. Put in garlic and red chilli. Toss well. Season with sugar and salt. Put in the fried sweet corn and toss well. Sprinkle with the five-spice salt from step 3. Toss again to coat evenly. Put in spring onion. Toss well and serve.

阿爺秘技 —————— Cooking tips

- 切粟米件時要連着少許芯，否則粟米粒會散開來。
- 粟米先蒸至大半熟透，拉油後令外皮香脆、內裏軟糯。
- 椒鹽的比例是 1 份五香粉：2 份鹽。
- 下椒鹽時要離火，如未能好好掌握，可將粟米先盛起才灑入椒鹽，因鑊裏仍有餘溫，會令五香粉有燶味。

- When you slice the sweet corn, make sure you keep part of the cob. Otherwise, the kernels will come apart.

- The sweet corn is steamed until medium-well done. Then it is fried in oil to make it crispy on the outside and chewy on the inside.

- To make the five-spice salt, the ratio between five-spice powder and salt is 1 to 2 by volume.

- Before you add five-spice salt to the sweet corn, turn the heat off. If you're not sure whether you can toss the sweet corn quickly enough in the wok, you may transfer the sweet corn onto a serving plate first, and sprinkle the five-spice salt over instead. The wok is still hot after the heat is turned off. If you're not stirring quickly enough, the five-spice salt may burn and smell bad.

百花釀甜豆

~Sugar snap peas stuffed with minced shrimp~

菜式特點

客家人早期居住在中原地區，以吃餃子為主，後來南移之後，沒有足夠的小麥做餃子皮，他們將豆腐切得很薄，用來包餡料。他們也會用其他蔬菜如苦瓜等釀料，慢慢形成了釀菜的文化。這道「百花釀甜豆」的工夫繁多，現已少人製作了。

材料

甜豆_____	10-12 條
蝦仁_____	200 克
蛋白_____	1 個
蝦醬_____	2 湯匙

Ingredients

- 10 to 12 pods sugar snap peas
- 200 g shelled shrimps
- 1 egg white
- 2 tbsp fermented shrimp paste

調味料

糖_____	1 茶匙
鹽_____	半茶匙
胡椒粉_____	少許

Seasoning

- 1 tsp sugar
- 1/2 tsp salt
- ground white pepper

阿爺秘技 ——————— Cooking tips

- 蛋白令蝦膠更滑、更香、更爽；將蝦膠和蛋白拌勻後，再冷藏 30 分鐘，令效果更佳。
- 將甜豆粒與蝦膠混和，再釀入甜豆莢內，令蝦膠有甜豆的清新味道。
- Adding egg white to the minced shrimp make it silkier, springier and more aromatic. Refrigerate the mixture for 30 minutes so that the minced shrimp can pick up the egg white nicely.
- I put the peas back in with the minced shrimp for a balance in flavour.

做法

1..... 蝦洗淨，用毛巾捲好，冷藏 30 分鐘吸乾水分，拍打成蝦膠。
2..... 蝦膠加入蛋白拌勻，下糖、鹽、胡椒粉調味，冷藏 30 分鐘。
3..... 甜豆洗淨、撕去筋，切開豆莢一邊，取出豆粒。
4..... 煮沸水，下糖及油，分別加入豆粒及甜豆莢稍灼至四成熟。
5..... 甜豆粒和蝦膠混和，釀入甜豆莢內，撲上薄薄一層生粉。
6..... 起油鑊，用慢火煎香蝦膠甜豆至金黃色，上碟。
7..... 蝦醬加少許糖煮至糖溶化，蘸百花釀甜豆享用，鮮味更突出。

Method

1. Rinse the shrimps and roll them in a clean towel. Refrigerate for 30 minutes to dry them up. Crush and chop into minced shrimp.

2. Add egg white to the minced shrimp. Then season with sugar, salt and ground white pepper. Refrigerate for 30 minutes.

3. Rinse sugar snap peas and tear off the tough veins. Cut off the straight side of each pod. Open the pod to remove all peas inside. Set the peas and the empty pods aside.

4. Boil water in a pot. Add sugar and oil. Blanch the pods and the peas separately until half-cooked.

5. Put the peas into the minced shrimp mixture. Mix well and stuff the empty pods with the mixture. Coat them in potato starch lightly.

6. Heat wok and add oil. Shallow-fry the stuffed sugar snap peas over low heat until golden. Save on a serving plate.

7. In a small pan, heat fermented shrimp paste with a dash of sugar. Cook until sugar dissolves. Serve it as a dip on the side with the stuffed sugar snap peas to accentuate the umami.

豆香餅

~Soybean dregs pancakes~

菜式特點

以前的生活條件不太好，人們會將磨豆漿過濾後剩下的豆渣做成各款菜式，不會浪費任何食物！

材料

豆渣＿＿＿＿＿＿150 克
韭菜＿＿＿＿＿＿100 克
麵粉＿＿＿＿＿＿180 克
鴨蛋＿＿＿＿＿＿2 個
水＿＿＿＿＿＿＿300 毫升

Ingredients

- 150 g soybean dregs
- 100 g Chinese chives
- 180 g plain flour
- 2 duck eggs
- 300 ml water

調味料

糖＿＿＿＿＿＿＿1 茶匙
鹽＿＿＿＿＿＿＿1 茶匙

Seasoning

- 1 tsp sugar
- 1 tsp salt

阿爺秘技

- 豆渣不要浪費，也可炮製成多款營養豐富的菜式，如製成肉餅、湯圓等。
- 鴨蛋比雞蛋濃香，而且韌性高；在中醫角度來説更有補氣的功效。

Cooking tips

豆渣 Soybean dregs

- After making soymilk, do not throw away the soybean dregs. Use them to make an array of nutritious dishes, such as meat patties or glutinous rice balls.
- Duck eggs have stronger flavours and more resilient texture than chicken eggs. From Chinese medical point of view, duck eggs are believed to boost Qi (vital energy) in the body.

做法

1...... 韭菜切碎;鴨蛋拂勻。
2...... 麵粉、水、鴨蛋液、豆渣、韭菜碎拌勻,灑入糖、鹽調味。
3...... 起油鑊,下粉漿煎成餅狀,反轉煎至兩面金黃色即完成。

Method

1. Finely chop the Chinese chives. Whisk the duck eggs.
2. In a bowl, mix flour, water, duck eggs, soybean dregs and Chinese chives together. Season with sugar and salt.
3. Heat a pan and add oil. Pour in the batter and fry into round patties. Flip them to fry until both sides golden. Serve.

豆乳浸時蔬

~Leafy greens poached in soymilk~

菜式特點

以前，鶴藪白菜被稱為「矮腳白菜」，又矮又胖，梗部很寬，質軟味甜，曾經大受菜販及酒樓歡迎。這道「豆乳浸時蔬」由於豆漿甜而不夠鮮味，所以加入花蛤取其鮮味，令口感更豐富。

材料

豆漿＿＿＿＿＿＿2 公升
花蛤＿＿＿＿＿＿300 克
鶴藪白菜＿＿＿＿500 克
薑＿＿＿＿＿＿＿4-5 片

調味料

糖＿＿＿＿＿＿＿1 茶匙
鹽＿＿＿＿＿＿＿1 茶匙

Ingredients

- 2 litres soymilk
- 300 g clams (Venus shells)
- 500 g "Hok Tau" Bok Choy
- 4 to 5 slices ginger

Seasoning

- 1 tsp sugar
- 1 tsp salt

鶴藪白菜

阿爺秘技 — Cooking tips

- 煮豆乳湯必須注意火候，溫火較適宜。
- 先放花蛤煮，令豆漿滲有海鮮的鮮味，隨後下鶴藪白菜吸收鮮甜之味。
- 煮菜不要加蓋，否則菜葉會變黃。
- Make sure you watch the heat when cooking soymilk. It's advisable to cook it over medium-low heat.
- The clams impart a seafood umami into the soymilk. The Bok Choy would pick up that sweetness and seafood flavours later.
- When you cook vegetables, do not cover the lid. Otherwise, the leaves would turn yellow.

做法

1..... 花蛤汆水至開口，撈起隔水，去掉沒開口的
花蛤。

2..... 鶴藪白菜洗淨，瀝乾備用。

3..... 豆漿煮沸，加入花蛤，下糖、鹽、薑片，放
入鶴藪白菜，煮至白菜莖呈半透明即成。

Method

1. Blanch the clams until they just start to open.
   Drain and set aside. Discard any clam that
   doesn't open.
2. Rinse the Bok Choy and drain well.
3. Bring soymilk to the boil. Put in the clams.
   Add sugar, salt, ginger slices and the Bok
   Choy. Cook until the Bok Choy stems turn
   translucent. Serve.

豆腐煎蛋

~Fried tofu omelette~

菜式特點

雖然只是兩款簡單的食材——豆腐及雞蛋,用熟練的刀工將豆腐切成薄片,粗菜精做,炮製一款令人耀眼的豆腐佳餚。

材料

硬豆腐＿＿＿＿＿＿＿2 件
雞蛋＿＿＿＿＿＿＿2 個
紅甜椒＿＿＿＿＿＿20 克（切粒）
青甜椒＿＿＿＿＿＿20 克（切粒）
蒜蓉＿＿＿＿＿＿＿10 克
乾葱蓉＿＿＿＿＿＿5 克
生粉＿＿＿＿＿＿＿1 茶匙

調味料

蠔油＿＿＿＿＿＿＿25 克
糖＿＿＿＿＿＿＿少許
鹽＿＿＿＿＿＿＿少許
水＿＿＿＿＿＿＿100-120 毫升

Ingredients

- 2 cubes firm tofu
- 2 eggs
- 20 g red bell pepper (diced)
- 20 g green bell pepper (diced)
- 10 g grated garlic
- 5 g chopped shallot
- 1 tsp potato starch

Seasoning

- 25 g oyster sauce
- sugar
- salt
- 100 to 120 ml water

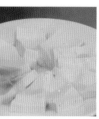

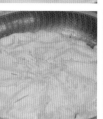

做法

1. ..... 豆腐用鹽水浸約 1 小時至硬身。豆腐切片，每片約 1 厘米厚，吸乾水分。

2. ..... 先在碟內加入少許蛋液，長方形豆腐片排於圓形碟內（碟的直徑最好比平底鑊略少，碟可嵌入平底鑊為適合），豆腐互相重疊，圓心位置成一小井口。

3. ..... 雞蛋拂勻，於井口倒入部分蛋液，慢慢由中心向外流，蛋液成為豆腐片和碟之間的潤滑劑。

4. ..... 平底鑊熱油，小心把（3）從碟內整份滑入平底鑊，保持豆腐成圓形的形狀，再倒入半份蛋液，讓蛋液均勻鋪滿平底鑊，下鹽調味，煎一面至定型後，小心把蛋餅溜回至碟上。

5. ..... 將平底鑊反轉蓋於碟上，鑊和碟同時一翻，蛋餅另一面可安全回鑊，倒入餘下蛋液，把蛋餅煎至金黃，上碟。

6. ..... 生粉 1 茶匙加入水調勻成生粉水，備用。

7. ..... 小鍋內加入水 100-120 毫升，下蠔油煮滾，加入乾葱蓉、蒜蓉、青紅甜椒粒，灑入糖、鹽調味，煮沸後加入生粉水勾芡，淋於豆腐煎蛋上即成。

阿爺秘技 — Cooking tips

- 豆腐浸泡鹽水後，質地較挺身，不容易散開來。
- 用適合尺寸的碟輔助翻轉豆腐，有助保持美觀的形狀。
- After soaking the tofu in salted water, it turns firmer in texture and is less likely to break down into bits.
- Pick a plate or dish of the right size to help flip the omelette. It's easier to maintain its shape that way.

Method

1. Soak the tofu in salted water for about 1 hour to stiffen it up. Cut tofu into slices about 1 cm thick. Wipe dry.

2. Pour some whisked egg into a dish. Then put the tofu slices over the whisked egg in the dish, fanning them out so that each slice overlaps with the next in a circular form with a well at the centre. (It's advisable to use a dish slightly smaller than your pan, so that it can sink right into the pan.)

3. Whisk the eggs. Pour some whisked egg into the well at the centre of the fanned tofu. Let it seep outward slowly. The whisked egg would act as lubricate so that the tofu can be slid down the pan easily.

4. Heat pan and add oil. Carefully slide the tofu from the dish into the pan while keeping its circular form. Then pour in half of the remaining whisked egg to fill the pan. Season with salt. Fry until one side is set. Carefully slide the omelette back in the dish.

5. Put the pan upside over the dish. Flip both the pan and the dish simultaneously to put the omelette back in with the fried side up. Pour in the rest of the whisked egg. Fry until golden. Save on a serving plate.

6. Mix 1 tsp of potato starch with some water. This is the potato starch slurry.

7. Pour 100 to 120 ml of water into a small pot. Add oyster sauce and bring to the boil. Put in shallot, garlic, and diced bell peppers. Season with sugar and salt. Bring to the boil and stir in the potato starch slurry from step 6. Cook till it thickens. Dribble over the tofu omelette. Serve.

金蒜番茄炒蛋

~Scrambled egg with tomato and fried garlic~

菜式特點

看似家常菜的番茄炒蛋，加點花樣，點睛之處竟是炸香
的金蒜，利用金蒜之香氣配搭甜美的番茄，香濃可口，
令人有耳目一新之感覺。

材料

番茄＿＿＿＿＿＿300 克（2 大個）
雞蛋＿＿＿＿＿＿4 個
蒜頭＿＿＿＿＿＿2 粒
葱＿＿＿＿＿＿＿10 克
糖＿＿＿＿＿＿＿2 茶匙
鹽＿＿＿＿＿＿＿1 茶匙

Ingredients

- 300 g tomatoes (2 large ones)
- 4 eggs
- 2 cloves garlic
- 10 g spring onion
- 2 tsp sugar
- 1 tsp salt

做法

1..... 蒜頭切粒備用。
2..... 番茄去皮、切小件備用。
3..... 葱切成葱花，備用。
4..... 雞蛋拌開（不用拌勻），加鹽調味。
5..... 鑊內加油，下蒜粒炸炒至金黃色，放入番茄炒勻，下糖調味，盛起備用。
6..... 在原鑊內加油，下蛋液炒至半熟，加入番茄炒勻至八成熟，熄火，灑上葱花即可。

Method

1. Finely dice the garlic.
2. Peel the tomatoes. Cut into small pieces.
3. Finely chop spring onion.
4. Whisk the eggs (without whisking too well). Season with salt.
5. Heat wok and add oil. Deep-fry the garlic until golden. Put in the tomatoes. Toss well and season with sugar. Set aside.
6. In the same wok, add oil. Pour in the whisked eggs and cook while stirring until half-set. Put in the tomato and cook until medium-well done. Turn off the heat. Save on a serving plate. Sprinkle with finely chopped spring onion on top. Serve.

阿爺秘技 / Cooking tips

- 利用蒜香提升番茄炒蛋之味道。
- 以餘溫炒蛋，可保持雞蛋嫩滑的口感。
- Adding garlic to the classic recipe elevate the taste profile to the next level.
- I turn off the heat before the eggs achieve desired doneness and let the gentle residual heat to cook them through. That's the key to velvety scrambled egg.

清心丸綠豆爽

~Split mung bean sweet soup with potato starch cubes~

菜式特點

清心丸是潮汕地區的特有小吃，具有清熱消暑的作用。
今次教你在家自製「清心丸綠豆爽」，軟滑清甜，加入
蓮藕粉水帶出之香氣，是夏日的必吃甜品。

材料

開邊綠豆＿＿＿＿＿50 克
冰糖＿＿＿＿＿＿＿80 克
蓮藕粉＿＿＿＿＿＿2 湯匙
水＿＿＿＿＿＿＿＿2 公升

清心丸材料

馬鈴薯粉＿＿＿＿＿150 克
暖水（70℃）＿＿＿150 毫升
熱滾水＿＿＿＿＿＿300 毫升

Ingredients

- 50 g split mung beans
- 80 g rock sugar
- 2 tbsp lotus root starch
- 2 litres water

Potato starch cubes

- 150 g potato starch
- 150 ml warm water (70°C)
- 300 ml boiling hot water

蓮藕粉 Lotus root starch

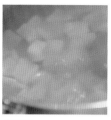

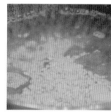

⊙ 示範短片

做法

1...... 開邊綠豆用水浸約 1 小時，隔水蒸約 20 分鐘。

2...... 馬鈴薯粉加入 70℃暖水拌勻。

3...... 接着加入熱滾水，一邊加入一邊攪拌，注意粉漿會在剎那間凝固，放涼後取出，切粒成清心丸。

4...... 將清心丸放入沸水煮約 5 分鐘至透明。

5...... 蓮藕粉加入 4 湯匙冷開水拌勻，再與少許熱水拌勻。

6...... 煮沸水 2 公升，加入開邊綠豆及冰糖，煮至冰糖融化，加入清心丸，邊攪拌邊加入蓮藕粉水，完成。

Method

1. Soak the split mung beans in water for 1 hour. Steam for 20 minutes.

2. Put potato starch into a bowl. Pour in warm water. Mix well.

3. Then plunge in boiling hot water while stirring the mixture continuously. The batter would set all of a sudden. Leave it to cool and turn it out of the bowl. Dice into potato starch cubes.

4. Blanch the potato starch cubes in boiling water for about 5 minutes until they turn transparent.

5. Put lotus root starch in a bowl. Add 4 tbsp of cold drinking water. Mix well. Pour in some hot water and stir again.

6. Boil 2 litres of water in a pot. Put in the split mung beans and rock sugar. Cook until sugar dissolves. Add potato starch cubes from step 4. Stir in the lotus root starch slurry from step 5. Serve.

阿爺秘技 ——————— Cooking tips

- 用熱滾水拌馬鈴薯粉時，熱滾水的份量及熱度是清心丸凝固的關鍵。
- 清心丸蘸上薄薄的乾粉，冷藏可保存數天，煮前取出即可。
- 最後加入蓮藕粉水，令糖水帶蓮藕之香氣，濃稠之餘，口感更滑溜。

■ To make the potato starch cubes successfully, the volume and temperature of the boiling hot water are the key.

■ After the potato starch cubes are blanched, you can coat them lightly in potato starch and refrigerate them. They last in the fridge for a few days. Just take them out of the fridge and add them to the sweet soup.

■ I stir in lotus root starch slurry at last to impart a mild lotus root fragrance and to thicken the sweet soup. It tastes more velvety this way.

明火煮食，鑊氣十足，煤氣煮食爐火力特強，
能均勻加熱整個鍋身，防止營養流失，
讓您輕易炮製不同美味佳餚，一同享受
明火煮食的樂趣！

明火煮食

樂趣多

煤氣
Towngas

The Green Choice
TGC

超強火力

TGC 極炎火嵌入式平面爐
MEGA2

🔥 6.0 千瓦爐頭，火力特強

🕐 預校熄火時間功能，煮食更簡易

▨ 德國頂級陶瓷玻璃

時尚色彩

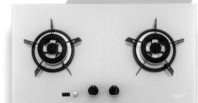

TGC 密封爐頭嵌入式平面爐
TRTB62ST-G

🎨 多種顏色選擇，配合時尚家居

▨ 密封式爐頭設計，方便清潔

◉ 火力達 5.0 千瓦，兼備獨立芯火

煮飯必備

TGC 煮飯寶
RJ3R

⊖ 內置煮飯功能

🔥 6.6 千瓦特大火力炒鑊爐頭

著者
李家鼎

策劃
謝妙華

責任編輯
簡詠怡、譚麗琴、陳芷欣

資料審閱
馬嘉茵

資料搜集及撰稿
吳紫玲、嚴寶麗

翻譯
Wendell J. Leers

裝幀設計
鍾啟善

攝影
《阿爺廚房》製作組

排版
何秋雲

出版者
萬里機構出版有限公司
香港北角英皇道 499 號北角工業大廈 20 樓
電話：(852) 2564 7511
傳真：(852) 2565 5539
電郵：info@wanlibk.com
網址：http://www.wanlibk.com
　　　http://www.facebook.com/wanlibk

發行者
香港聯合書刊物流有限公司
香港荃灣德士古道 220-248 號荃灣工業中心 16 樓
電話：(852) 2150 2100
傳真：(852) 2407 3062
電郵：info@suplogistics.com.hk
網址：http://www.suplogistics.com.hk

承印者
美雅印刷製本有限公司
香港九龍觀塘榮業街 6 號海濱工業大廈 4 樓 A 室

規格
特 16 開（240mm x 170mm）

出版日期
二〇二〇年十二月第一次印刷

本書由電視廣播有限公司授權出版